Science in the Metropolis

This book presents new research on spaces for science and processes of interurban and transnational knowledge transfer and exchange in the imperial metropolis of Vienna in the late nineteenth and early twentieth centuries. Chapters discuss Habsburg science policy, metropolitan natural history museums, large technical projects including the Ringstrasse and water pipelines from the Alps, urban geology, geography, public reports on polar exploration, exchanges of ethnographic objects, popular scientific societies, and scientifically oriented adult education. The infrastructures and knowledge spaces described here were preconditions for the explosion of creativity known as 'Vienna 1900'.

Mitchell G. Ash is Professor Emeritus of Modern History at the University of Vienna, Austria, and a member of the Berlin-Brandenburg Academy of Sciences and Humanities as well as the European Academy of Sciences and Arts.

Routledge Studies in Cultural History

The Humanities in Transition from Postmodernism into the Digital Age
Nigel A. Raab

Negotiating Memory from the Romans to the Twenty-First Century
Damnatio Memoriae
*Edited by Øivind Fuglerud, Kjersti Larsen, and
Marina Prusac-Lindhagen*

Cultures and Practices of Coexistence from the Thirteenth Through
the Seventeenth Centuries
Multi-Ethnic Cities in the Mediterranean World, Volume 1
Edited by Marco Folin and Antonio Musarra

Controversial Heritage and Divided Memories from the
Nineteenth Through the Twentieth Centuries
Multi-Ethnic Cities in the Mediterranean World, Volume 2
Edited by Marco Folin and Heleni Porfyriou

History as Performance
Political Movements in Galicia Around 1900
Dietlind Hüchtker

The Cultural Life of Risk and Innovation
Imagining New Markets from the Seventeenth Century to the Present
Edited by Chia Yin Hsu, Thomas M. Luckett, and Erika Vause

Popular New Orleans
The Crescent City in Periodicals, Theme Parks, and Opera, 1875–2015
Florian Freitag

Science in the Metropolis
Vienna in Transnational Context, 1848–1918
Edited by Mitchell G. Ash

For more information about this series, please visit: www.routledge.com/
Routledge-Studies-in-Cultural-History/book-series/SE0367

Science in the Metropolis

Vienna in Transnational Context, 1848–1918

Edited by Mitchell G. Ash

Routledge
Taylor & Francis Group

NEW YORK AND LONDON

First published 2021
by Routledge
52 Vanderbilt Avenue, New York, NY 10017

and by Routledge
2 Park Square, Milton Park, Abingdon, Oxon, OX14 4RN

Routledge is an imprint of the Taylor & Francis Group, an informa business

© 2021 Taylor & Francis

Library of Congress Cataloging-in-Publication Data
Names: Ash, Mitchell G., editor.
Title: Science in the Metropolis : Vienna in transnational context,
   1848–1918 / edited by Mitchell G. Ash.
Description: 1st. | New York, NY : Routledge, 2021. | Series: Routledge
   studies in cultural history ; 96 | Includes bibliographical references
   and index.
Identifiers: LCCN 2020023768 (print) | LCCN 2020023769 (ebook) |
   ISBN 9780367612580 (hardback) | ISBN 9781003104865 (ebook) |
   ISBN 9781000210217 (adobe pdf) | ISBN 9781000210224 (mobi) |
   ISBN 9781000210231 (epub)
Subjects: LCSH: Science—Austria—Vienna—History—19th century. |
   Science—Austria—Vienna—History—20th century. | Scientific
   ability—Austria—Vienna—History—19th century. | Scientific
   ability—Austria—Vienna—History—20th century. | Science and
   state—Austria—Vienna—History—19th century. | Science and
   state—Austria—Vienna—History—20th century. | Vienna
   (Austria)—Intellectual life. | Vienna (Austria)—Civilization. |
   Infrastructure (Economics)
Classification: LCC Q127.A8 S2525 2021 (print) | LCC Q127.A8 (ebook) |
   DDC 509.436/1309034—dc23
LC record available at https://lccn.loc.gov/2020023768
LC ebook record available at https://lccn.loc.gov/2020023769

ISBN: 978-0-367-61258-0 (hbk)
ISBN: 978-1-003-10486-5 (ebk)

Typeset in Sabon
by Apex CoVantage, LLC

Contents

Figures

Tables

Acknowledgments

The chapters in this volume are revised and expanded versions of papers presented at the conference "Science in the Metropolis: Places and Constellations of Scientific Knowledge 1848–1918," held at the Austrian Academy of Sciences 16–18 November 2017, and supported by the Commission for History and Philosophy of Science of the Austrian Academy of Sciences.

1 Metropolitan Scientific Infrastructures and Spaces of Knowledge in Vienna, 1848–1918

An Introduction

Mitchell G. Ash

1. Why Vienna—and What Are Metropoles?

'Science and the City' (Dierig et al. 2003; Hochadel and Nieto-Galán 2018) has become an important topic of research in recent years, for good reasons. That urban settings have been central to the emergence and growth of modern science, and that science and technology have had formative significance for urban environments, are well-accepted statements. How urbanization has made possible the intensification of scientific networks, and how research institutions such as universities, observatories, and museums have functioned as public ornaments of great as well as smaller cities are well-researched topics for a number of locations and time periods. Less well researched until quite recently, however, is the case of Vienna, the capital city, cultural metropolis, and scientific center of the Habsburg Empire.

The significance of rapid urbanization in Vienna after 1848 for the complex political, social, and cultural conflicts of the late Habsburg monarchy is well studied, and Viennese modernism, or 'Vienna 1900', has been an intensively researched and debated topic since the pioneering studies of Carl Schorske (Schorske 1980; Beller 2001). Recent scholarship has shown that not only Freudian psychoanalysis, but numerous other advances in the sciences and medicine were part and parcel of the culture of 'Vienna 1900' (for discussion and literature see Kandel 2012; Ash 2018). Examples include *inter alia* the theoretical physics of Ludwig Boltzmann (Reiter 2017), the philosophy of science of Ernst Mach (Stadler 2019), the probabilistic physics of Franz Serafin Exner (Coen 2007), and Karl Landauer's discovery of blood groups (Kandel 2012), all at the University of Vienna, studies of radioactivity at the Radium Institute of the Academy of Sciences (Reiter 2017), and experimental biology at the *Biologische Versuchsanstalt*, also called the Vivarium (Müller 2017, Sienell et al. 2016), some of which attracted public scandal, such as Eugen Steinach's studies on sexuality and sexual hormones (Walch 2016) and

Paul Kammerer's work on inheritance, now regarded as a step toward modern epigenetics (Taschwer 2019).

In *Vienna in the Age of Uncertainty*, Deborah Coen (2007) successfully challenged Schorske's thesis that the defeat of political liberalism and the rise of political antisemitism in Vienna in the late nineteenth century led to an era of illiberal cultural pessimism. Instead, we might rather speak of a dynamic tension between a liberal progressive spirit based on a clear-eyed acceptance of probabilistic thinking that, as Coen shows, was still very much alive in 1900, especially though not only in the natural sciences, and reactionary resistance to modernism, motivated in large part by antisemitism, especially though not only in medicine (Buklijas 2012; Seebacher 2010). All of this and more should be included when the by now well-accepted claim is made that the legendary artistic creativity and intellectual innovations of 'Vienna 1900' were linked to a high degree with "the cultural infrastructure of the big city" (Lenger 2013, 243).

However, the role of the sciences and technology in establishing what could be called the *infrastructural preconditions of Viennese modernism* long before 1900 is less well recognized, beyond what Schorske wrote so well about the Ringstrasse decades ago (for an early statement of this point see Ash 1999). The case of Vienna should be of obvious importance for metropolitan studies as well as the history of science and technology, given the physical nearness of an emerging research university, newly founded or restructured scientific technical schools, and a number of internationally recognized extra-university research institutes both to one another and to the state institutions that supported them (see Section 2). The central question of this volume, therefore, is how this shared metropolitan space made possible new urban, interurban, regional, and transnational linkages of academic and non-academic researchers, along with new efforts to make the sciences accessible to wider publics in the emerging Habsburg metropolis in the late nineteenth and early twentieth centuries, and how these dynamics became interlinked with metropolitan science and technology in other European and North American centers. The goals of the volume are: (1) to contribute to and further stimulate the growing historiography of the sciences and technology during the late Habsburg Empire; and at the same time (2) to benefit from and also advance metropolitan studies by continuing the long-running discussion of whether and how to distinguish metropoles from other large cities in historical context.

Why should we focus on this period, or on Vienna, and why should we continue to be so Eurocentric? After all, public discussion on urban spaces turned away some time ago from Europe to the so-called mega-cities of Asia, Africa, and Latin America (see *inter alia* Stratman 2011; Kraas 2014). Places like Beijing, Shanghai, Tokyo, Mumbai, or Sao Paulo make Vienna and even Berlin seem medium-sized by comparison. Of the Euro-American metropoles, only London and New York can compete any longer with these sprawling giants, size-wise. One Asian 'megacity', Tokyo, had already become a metropolis in the nineteenth century. Nonetheless, the last third of the nineteenth century remains the period of the most rapid and widespread urbanization in European history until that time, which also saw the emergence of new and the rapid growth of existing metropoles in Central and Eastern Europe, including Vienna, Berlin, and St. Petersburg, that soon rivaled and became ever more intensively linked with their older counterparts London and Paris.

Two of these new metropoles will interest us here: Berlin, capital of Prussia and from 1871 of the German Empire, the population of which grew by more than 900 percent, from 437,000 (fifth among European cities) to 4 million (third place) between 1850 and 1913 (Lenger 2013, 53), although it did not achieve its full geographical extension until 1920 (for an overview see Hessler 2010); and above all Vienna, long-time seat of the Habsburg monarchy, the population of which also grew rapidly when it incorporated dozens of suburbs in 1857, from 431,000 (sixth among European cities) in 1850 to 2,150,000 (fifth place) in 1913 (Lenger 2013, 53), but which achieved political status as a self-governing entity separate from Lower Austria only in 1920. Of course we could also add New York and Chicago to this mix, or Budapest, from 1867 the second capital of the Dual Monarchy and an important center of scientific and scholarly culture its own right (Nernes 2010; Gantner 2017; Stráner 2018). The point I want to emphasize here is already implied in the word 'emergence'. Metropoles are by no means only modern phenomena; Nineveh and Babylon should tell us that, and these examples also show that metropoles can come to be and also pass away. Metropoles have always been moving targets, and this became unmistakably clear, at least in Europe, during the period discussed here.

Of course, other large cities grew rapidly during this period, in Europe and elsewhere, some of them still more so than the metropoles. Places like Essen, the home of Krupp Steel, only became large cities at all at this time. Already important port cities like Barcelona, Hamburg, and Rotterdam acquired new significance, and also scientific and cultural infrastructures, in this period (on Barcelona see Hochadel and Nieto-Galán 2016). Thus it is justified to ask why this volume is entitled 'Science in the Metropolis'. What distinguishes metropoles from other lage cities? Precisely because metropoles are and have always been moving targets, it may seem futile to offer a fixed, universally valid definition of this concept. Current metropolitan studies assign the term no more than historical meaning as an actor's category in any case, and some scholars reject the term 'metropolis' itself as an overused marketing tool (Brantz et al. 2012, 11). For historical research, however, it remains important to ask which actors in the past employed the term in what circumstances and why. Perhaps it is therefore permissible to offer some elements of such a definition, in order to begin the discussion.

What defined a metropolis in the late nineteenth century, I wish to suggest (following Reif 2012, 32–33), is a unique concentration of actual political power and economic, financial, and cultural predominance, combined with subjectively ascribed cultural cachet; all of this, taken together, made such cites "reference points" and "role models" (Mieg 2010, 322), indeed "a locus of dreams" for many (Brantz et al. 2012, 14; see also Brantz, Chapter 2 in this volume), and at the same time places of extreme poverty and labor exploitation. Most, though not all, of the great cities just named as nineteenth-century European metropoles were capital cities of existing empires or emerging nation-states. Each was also a center of economic activity, much of which was and had long been commercial due to high demand for luxury goods from the rulers and their courts. Added to this in the period discussed here was the growth of light and heavy industry, for which firms like Borsig and Siemens stood in Berlin (Wise 2018). This was also true to some extent for Vienna in the late nineteenth century, though the chemical industry was located mainly in Bohemia and not in the capital (Rosner 2004). The role of metropoles as financial hubs was exemplified at first positively in Vienna by the commercial boosterism that promoted the holding of the World Exhibition there in 1873 (*inter alia* to mark the twenty-fifth anniversary of the Emperor's coronation), and negatively by the stock market crash of that year, which began in New York but soon reached Vienna and along with a cholera outbreak, led to the fair's being perceived as a disaster (Kos and Gleis 2014).

The dimension of cultural power is embodied in Vienna by the Imperial and Royal Court Theater (now the Burgtheater) and the Imperial and Royal Court Opera House (*k. und k. Hofoperntheater—Neues Haus*, now the Vienna State Opera), opened in 1869, as well as the challenges to their standing in this period by the establishment of opera houses in Budapest and elsewhere in the empire (Prokopovych 2014) and so-called people's theatres and opera houses (*Volkstheater* and *Volksoper*) in Vienna (opened in 1889 and 1898 respectively) and elsewhere (Ther 2012). As Marianne Klemun points out (in Chapter 5 of this volume), the term 'metropolis' first appeared in Vienna newspapers in the 1860s, with explicit reference to the city as a center of "taste". As Elias Canetti wrote in a memoir, referring to life in his childhood home of Rutschuk, in Bulgaria: "When someone journeyed up the Danube toward Vienna, people said 'he's going to Europe'" (quoted in Lenger 2013, 50. Translation MGA).

How did the sciences and technology factor into this emerging metropolitan mix? I referred previously to the infrastructural foundations of Viennese modernism. Speaking in this way implies taking a particular approach to this topic, one that emphasizes networks of power, technology, and knowledge. Such infrastructures can take a variety of forms. One of these emphasizes technical infrastructure of the kind involved in the planning, design, and construction of metropoles, including boulevards and canals, street lighting, sewer systems, water and energy supplies, as Sándor Békési does in this volume (Chapter 4). Seen in this way, metropoles, like all large cities, can be described as networked collections of large technical systems in the sense elaborated long ago by Thomas Hughes (Hughes 1983; Mayntz and Hughes 1988). We should remember that such systems were, and remain, far more than modern conveniences. Two examples illustrate their transformative impact and their role in the constitution of modernity itself: the streetcar, at first horse-drawn and then electric, helped to overcome the boundary between the city and the countryside, while the bicycle made self-propelled mobility possible even for the less well-off (Oldenziel 2018); gas, and then electric lighting quite literally overcame the division between day and night in urban spaces, with profound impacts on both work and leisure time and spaces (Schivelbusch 1988). It is not at all far-fetched to suggest that such technical infrastructures were conditions of the possibility of modernity itself as a mode of living, and also as a style of thinking.

Once we begin to think about such issues in this way, linkages of scientists and scientific institutions with each of the defining elements of the metropolis just outlined become evident. The political dimension is of central importance in continental Europe, for the simple reason that scientific research institutions outside the industrial sector were generally funded by the state (for exceptions, see Section 2). In Vienna, the founding of the Imperial and Royal Academy of Sciences in 1847 (*Wiener Zeitung* 1847), shortly before the upheavals of 1848, preceded a series of measures taken by the restored imperial regime following the violent defeat of the revolution which, taken together, can well be understood as a Neo-Absolutist modernization project from above (on cultural politics in the 1850s and later, see Judson 2016, Chapter 6). The series began with far-reaching education reforms, including a restructuring of higher education, conceived by liberal philosopher Franz Exner (father of the previously mentioned physicist Franz Serafin Exner) during the revolution and carried out with important modifications by Minister of Religion and Education Count Leo Thun von Hohenstein from 1849 onward (Aichner and Mazohl 2017; Surman 2019, Chapter 2). However, although the replacement of cameralistic administration from above by limited faculty (meaning: professorial) self-governance was a central feature of this reform, professors continued to be appointed by the Emperor, as before. Thun paid personal attention to appointments in what he called the 'national' disciplines, including philosophy, history, law, and Slavic studies (Feichtinger 2010, Section 3.2).

It is widely acknowledged that the reforms of the Neo-Absolutist era made possible the emergence of research universities and science- and scholarship-based academic training throughout the empire, most notably in the imperial capital (on the reform of higher technical education in this period and later, see Mikoletzky 1995). Central loci of the growth and increasing differentiation of institutionalized research within the

Habsburg universities were the Philosophical Faculty, roughly equivalent to the faculty of arts and sciences in the United States, which incorporated both the natural sciences and the humanities (see for example Schübl 2010), and the Medical Faculty. By 1900 a total of 34 seminars, laboratories, institutes, and research or teaching collections had been established in the Philosophical Faculty of the University of Vienna; 30 disciplines were represented by a total of 68 full and associate professors as well as 70 (mainly unsalaried) instructors called *Privatdozenten* (Mühlberger 2009, 88. By this time, however, the rapid growth in teaching staff of earlier decades had largely ceased; see Ranzmaier 2015). Parallel to this was the increasingly research-oriented training at the Vienna Polytechnicum (founded in 1815), renamed Technische Hochschule in 1872, which received the right to award doctoral degrees in 1901 (Mikoletzky 2018), and at newer technical institutions, such as the school of agriculture and forestry (*Hochschule für Bodenkultur*), founded in 1872. As sites of science-based academic training and research, the lecture halls, laboratories, and seminars of the university and the higher technical schools were plainly basic components of the emerging scientific infrastructure of imperial Vienna, and therefore both signs and agents of what has come to be called the scientification of society (see for example Brückweh et al. 2012). However, they were by no means the whole story.

Equally significant for the natural sciences and historical scholarship were research institutions established outside the universities under direct state supervision, including the Imperial and Royal Geological Anstalt, founded in 1849 (Klemun 2012), the Imperial and Royal Central Anstalt for Meteorology and Earth Dynamics, founded in 1851 (Hammerl et al. 2001), and the Institute for Austrian Historical Research, founded in 1854 (Surman 2019, 72). These institutes were tasked respectively with organizing and executing empire-wide mapping and data-gathering surveys in geology, earthquake studies (Coen 2013), and weather and climate research (Coen 2018), or the collection of Medieval and later manuscripts in order to document the heritage of "Greater Austria". To these should be added the expedition of the frigate *Novara*, which circumnavigated the globe from 1857 to 1859 under the auspices of Archduke Ferdinand Maximillian, commander of the imperial navy, and brought back thousands of specimens later incorporated into the imperial collections (Kraus 2004). The common guiding idea for all of these efforts was the young Emperor Franz Josef I.'s watchword, '*Viribus unitis*', meaning "with united forces". Nearly all of these institutes' names began with the prefix 'Imperial and Royal', a direct reference to Franz Josef's positions as head of the empire and king of Hungary until 1867, and their task was nothing less than the knowledge-based construction of an empire that generally lacked natural borders as a cultural nation (*Kulturnation*) (Ash and Surman 2012, 12; Feichtinger 2012). Deborah Coen (2018, Chapter 4) elegantly summarizes the status and scope of these efforts and also the comprehensive, documentary research style involved with the sobriquet "imperial and royal science" (*k. und k. Wissenschaft*).[1]

Despite the evident primacy of the Emperor and the state he headed in the support of higher education and science, wealthy patrons, many of them Jewish, played important roles in the expanding infrastructure of scientific and scholarly research during this period. An early example is the banker Ignaz L. Lieben, whose son Adolf, a chemist, became the first Jewish full professor at the University of Vienna in 1875 (Ash 2013, 101). At his son's urging, Lieben endowed a prize awarded to outstanding young scientists beginning in 1865 (Soukup 2004; Reiter 2017, 71–116). Two later examples of privately supported, internationally recognized research in Vienna are: the Vivarium, the institute for experimental biology mentioned previously, founded in 1903 and financed personally by its three founders, Hans Przibram, Leopold von Portheim, and Wilhelm Figdor (Müller 2017); and the Radium Institute of the Academy of Sciences, also mentioned previously, founded in 1910 with the support of industrialist Karl Kupelwieser (Reiter 2001; Reiter 2017, 32–33, 199ff.). The participation of business people, professionals, and academics alongside liberal nobles in the creation and operation of a plethora of private scientific societies, such as the Association for the Dissemination of Natural Scientific Knowledge, founded in 1860, is well documented (Taschwer 1997). It should be emphasized, however, that the distinction between state and civil society was less clear in the Habsburg metropole than in the English-speaking world. Such societies or clubs were (and still are) required to register with the government, and many of those founded before 1867 even employed the prefix 'Imperial and Royal' in their names, indicating imperial connections (for further discussion see Johannes Mattes, Chapter 9 in this volume).

Links of the sciences and scholarship to metropolitan cultural life were most obviously visible in representative buildings such as the new main building of the University of Vienna, dedicated by the Emperor himself in 1884 (Mühlberger 2008; Rüdiger and Schweizer 2015, Part 2), and the Imperial and Royal Natural History Museum, begun in the 1860s and also opened by the Emperor in 1889. Both houses were built as part of the Ringstrasse ensemble; the Natural History Museum was located directly opposite the Art History Museum, and is the same size as its counterpart, thus reflecting the equal symbolic status of the empire's natural and artistic possessions (Jovanović-Kruspel and Schumacher 2017). Both museums were places of imperial representation, described then and later in the language of court society as the Emperor's gifts to his people, and at the same time scientific research facilities, thus integral components of the complex of 'imperial and royal' science and of Vienna's metropolitan scientific infrastructure (for further discussion see Brantz, Chapter 2 in this volume).

2.   Metropolitan Science—Nested Knowledge Spaces

The approach to this topic taken here focuses on metropolitan spaces and constellations of scientific knowledge in transnational contexts. Following the path marked in general history by Henri Lefebvre (1991) and others, the history of science has taken a 'spatial turn' in recent years (Ash 2000; Livingstone 2003; Livingstone and Withers 2011; Withers and Livingstone 2011). Of course, historians of science have long thought at least implicitly in spatial categories, for example when they have inquired into the complex processes by which locally situated knowledge acquired universal recognition (for older cases in point see Cohen 1987). As Oliver Hochadel and Ulrike Spring note (in Chapters 3 and 6 of this volume), the more recent 'spatial turn' is linked to the call for a global history of science, integrating the world-wide expansion of knowledge gathering primarily though not entirely under imperialist auspices, and the transnational communication and exchange of research ideas and resources. Examples of the former process in the Habsburg Empire are the *Novara* expedition and the empire-wide surveys begun during the Neo-Absolutist period, mentioned previously; examples of transnational communication and exchange will be considered in this volume (see the following paragraphs).

However, spatial perspectives can be employed to understand the co-evolution of the sciences and the Habsburg metropolis Vienna in a variety of ways. As a preliminary effort to structure the discussion, I propose to speak here of *physical, geographical/political, and social spaces of knowledge*. As the chapters in this volume make clear, none of these conceptualized 'spaces' is actually separate from the others; in reality all of them are nested in and intertwined, or, as Hochadel, Spring, and also Marianne Klemun (in Chapter 5 of this volume) put it, 'entangled' with one another. Common to all of them is the point that spaces are "not simply 'out there'," but are established performatively in speech acts, migration, and other human activities (Surman 2019, 5).

The category of *physical spaces* refers first of all to the buildings housing scientific research institutions, some of which have already been mentioned. In keeping with the historicist architectural style predominant at the time, many of these buildings (like others, such as the new City Hall, dedicated in 1883) were constructed in Vienna during the second third of the nineteenth century in Neo-Gothic style as secular cathedrals, imposing embodiments of long-standing tradition and also of the unity of beauty and truth (for the Natural History Museum as an example see Brantz, Chapter 2 in this volume, Figure 2.1). A second level of meaning within this category is embodied by the design of interior spaces, for example that of the geology hall in the Natural History Museum, which was (and still is) decorated with landscapes of the regions of the empire from which the exhibited stones and minerals came. When objects collected during the *Novara* and other expeditions were brought to such sites and classified, or explorers reported on the results of their travels in lecture halls, knowledge moved from one place to another and was transformed in the process.

This brings us to *geographical spaces*, which in turn can be considered at the micro level, meaning spaces within a single location, here: in the metropolis Vienna, at the meso level, meaning networking, circulation, and exchange regionally, between neighboring lands, or within the Habsburg Empire, and finally at the macro level, including transnational or global networking, circulation, and exchanges.

Considering geographical spaces at the micro level brings us to the concept (or metaphor) of 'thick space' formulated by Dorothee Brantz and her associates, in analogy to Clifford Geertz's concept of thick description (Brantz et al. 2012; see the also Brantz, Chapter 2 in this volume). This expression acquires real descriptive power when we consider closely linked ensembles of physical spaces of knowledge in the emerging imperial metropolis of Vienna. Vienna's new university building, the natural and art history museums, and the technical higher education institute (*Polytechnicum*) on the Karlsplatz, were all within walking distance of one another on or near the Ringstrasse. They were also within walking distance of the seat of Habsburg power, the Hofburg, as well as the seat of the Education Ministry which financed their work completely or in part, which was located from 1848 to 1861 in the Palais Rottal in the Singerstrasse (near St. Stephan's Cathedral), and from 1871 onward (until today) in the Palais Starhemberg on the Minoritenplatz, a stone's throw from the new university building. The Academy of Sciences had been awarded the former university building, an eighteenth-century palace in the southeast corner of the city center, in 1857, a location somewhat removed from the other buildings but still within walking distance. As the diary of geologist Franz von Hauer, director of the Imperial and Royal Geological Anstalt from 1866 to 1884 and then Ferdinand von Hochstetter's successor as *Intendant* (General Director) of the Natural History Museum, shows, Hauer traversed these 'thick spaces' of central Vienna regularly, as he circulated between his offices in the Geological Anstalt (in the Palais Rasumofsky in the third district) and later the Natural History Museum to the ministry or other relevant locations (Klemun (ed.) 2020).

Partly in connection with the extraordinary building activity during and following the construction of the Ringstrasse, this centralized microgeography changed in interesting ways during this period, and the residence patterns and related walking behavior of Vienna's academics changed with it. As Jan Surman (2020) has shown, in the 1850s university teachers lived mainly in the city center, near their work places (hospitals, churches, university buildings); by 1910 many of them had moved to addresses nearer to the new university main building. For professors of medicine, the relevant attractors were the laboratories and medical clinics in or near the city's General Hospital, located since the eighteenth century outside the city walls, after 1857 beyond the Ringstrasse in the Alsergrund (ninth district). Many instructors (*Privatdozenten*) of the Medical Faculty remained in the city center; Surman suggests that they did so because their income came from private practice rather than from teaching. Over time, as the city and its scientific infrastructure expanded and public transportation improved, the network of walkable 'thick spaces' became thinner, microgeographically speaking, extending to more distant institutions, such as the new university observatory, built in Währing (eighteenth district) between 1874 and 1879 and officially inaugurated by the Emperor in 1883, the Vivarium, mentioned previously, founded in 1905 and located in the Prater park (second district), and the spacious new buildings for the university institutes of physics and chemistry constructed along the Währinger Strasse (ninth district), also in the first decade of the twentieth century. As Johannes Mattes shows (in Chapter 9 of this volume, also see the following), the locations of scientific society meetings also spread out beyond the city center during this period.

As we move now to the meso level and consider exchanges within the empire, it becomes clear that the concept of 'knowledge spaces' also encompasses relationships within the empire, enacted for example in the empire-wide data-gathering and survey projects, named previously, but also in networks and academic migration between the metropolis and regional centers such as Prague and Budapest (Frank 2018; Svobodný 2018). As Jan Surman (2019, Chapter 4) has documented in detail, academic mobility and career patterns in this period nonetheless continued to be dominated by the metropolis. As he shows, the University of Vienna, the largest in the empire, was the primary supplier of graduates to the other universities in Cisleithania; roughly 40 percent of transfers by professors to these institutions between 1848 and 1918 came from Vienna, while transfers to Vienna from the other Cisleithanian universities in the same period totaled only 10 percent in the Medical and 13 percent in the Philosophical Faculty (ibid., 151, Table 6); of course many of these transfers to Vienna had been trained there in the first place. In sum, Surman writes, "other universities profited from graduates from Vienna, and Vienna could choose from among the best scholars from across the empire in its appointments"—a policy shared by the faculties and the ministry, which consistently acted to retain in Vienna or to bring in "the very best people" (ibid., 151, 155–156). Long-standing salary advantages, justified by Vienna's higher living costs, and higher incomes from student fees, which were eliminated only in 1898, reinforced metropolitan predominance.

However, as Oliver Hochadel notes (in Chapter 3 of this volume), circulation and exchange of scientific knowledge took place not only between the Viennese 'center' and the provincial 'periphery', but also among the regional 'centers' themselves (Gantner and Hein-Kircher 2017). Cities such as Cracow, Budapest, and Prague were also home to well-regarded universities and technical schools as well as Academies of Sciences, and thus presented themselves and also acted as scientific centers in their own right, becoming focal points of emerging Polish, Hungarian, and Czech national cultures, respectively (see, for example, Štrbáňová 2012). The emergence of increasingly assertive nationalist movements inhibited the free circulation of scientists and scholars within the empire, due to the desire of those native to these regions to remain or return there, and the pressure on outsiders to teach in the local languages (Surman 2012). Nonetheless, Vienna remained a primary reference point, not least because the imperial ministry remained responsible for professorial appointments in Cisleithania. Indeed, Surman (2019, 7) argues that nationalization in the provinces "began as a reaction to processes in Vienna, the intellectual center"; over time, nationalization in the provinces and intensified German nationalism in Vienna increasingly went hand in hand.

We move now from the regional toward the European and global levels. As Sándor Békési shows (in Chapter 4 of this volume), inter- and transnational dimensions of circulation and exchange were already evident in the creation of Vienna's modern technical infrastructure from the 1850s onward, as private contractors brought in from Britain, France, the German states, and elsewhere in Europe were involved in the construction of the famous drinking water pipeline from the Alps to the city, the regulation of the Danube, the construction of the Ringstrasse, and other large-scale projects. Explorers returning from the Arctic in the 1870s informed large numbers of interested citizens and debated with experts in Hamburg, Vienna, and elsewhere about their discoveries, creating what Ulrike Spring (in Chapter 6 of this volume) terms 'entanglements of knowledge'. As Brooke Penaloza-Patzak shows (in Chapter 8 of this volume), ethnographic research and exhibition materials circulated and were exchanged by cultural anthropologists amongst museums in Washington, DC, New York, Chicago, Berlin, and Vienna. To these examples we might add the transnational circulation of Radium extracted and purified from pitchblende ore mined in Bohemia, over which the empire held a practical monopoly, to the laboratories of Marie and Pierre Curie in Paris, or from Vienna to the laboratory of Ernest Rutherford in Manchester (Rentetzi 2008; Reiter 2017), and the transfer of live organisms for study from marine biological laboratories in Naples or Trieste to the Vienna Vivarium. Within the metropoles, individual scientific institutes within or outside university settings, for example the geographical institutes in Berlin and Vienna (as discussed by Svatek in Chapter 7 of this volume), themselves became nodes of national and supranational circulation networks, pulling in data and objects from far-flung locations in Europe and other continents for analysis and publication by researchers in each metropole.

Of course, not only objects and data, but also scientists and scholars themselves participated in such inter- and transnational circulation.

This brings us to *social or sociocultural space*. Relevant here are people whom Kapil Raj (2016), writing about knowledge circulation among non-European and European locations, has termed "go-betweens" or "cultural translators". Applied to the metropolis of Vienna, we can distinguish two sorts of "go-betweens": scientists and scholars moving from laboratories or study rooms to appointments with colleagues or officials in search of institutional power or influence, and aspiring or established scientists collaborating with interested amateurs at meetings of private scientific societies. The Habsburg Empire had long been and continued in this period to be an agglomeration of hierarchically structured societies dominated by large landholders and the military. During the second half of the nineteenth century, in some cases beginning much earlier, an educated bourgeoisie emerged and grew in Vienna, aided in part by the reforms in secondary, university, and technical education mentioned previously. Members of the educated middle classes increasingly sought alliances with liberally minded nobles, and vice versa, which enabled *inter alia* the establishment and maintenance of local, regional, and national scientific societies of various kinds. In addition to the Association for the Dissemination of Natural Scientific Knowledge, mentioned previously, an evocative example of such alliances is the first International Congress for Ornithology, held in Vienna in 1884 under the "protection and patronage" of Crown Prince Rudolf, a passionate student of birds, friend and bird-hunting partner of zoologist Alfred Edmund Brehm (Ursprung 1984; Holzleitner 2013, Chapter 6). The Society for Cave Study, discussed by Johannes Mattes (in Chapter 9 of this volume), is an example of how popular scientific societies overcame social and physical barriers at the same time, bringing together academics and amateurs above ground to converse about subterranean explorations.

Mattes also maps the changing physical locations of the meetings of such societies as spaces for scientific communication and exchange. When meetings took place in the university, the Natural History Museum, or the Academy of Sciences, they embodied the blurring of boundaries between the bourgeoisie and the imperial state already evident in their socially mixed membership and their legal status, mentioned previously. A new physical and social space for popular science supported by private funds was the Urania, founded in 1897; its new building in Art Nouveau design, with its own lecture hall and observatory located on the Danube canal, was dedicated by Archduke Karl Ferdinand in 1910. Alternative social spaces for scientific teaching, research, and discussion were created from the 1880s onward by the adult education movement, jointly supported by social democrats and liberal academics, and physically embodied in the *Volksheim* Ottakring, dedicated in 1905 (discussed in detail by Christian Stifter in Chapter 10 of this volume).

3. The Chapters

With that I come to the arrangement of this volume. Part One contains two historiographical overviews, opening with Dorothee Brantz's chapter entitled "Metropolitan Natural Histories: Inventing Science, Building Cities, and Displaying the World". Following a discussion of the 'metropolis' concept, Brantz takes a global approach, focusing on zoos and natural history museums as examples of how fast-growing nineteenth-century metropoles such as Berlin and Vienna brought the exotic natural world TO the city while at the same time representing that world IN the city. Oliver Hochadel's historiographical reflections on the urban history of science more broadly are a pendant to Brantz's analysis. Hochadel moves from work on science in an individual city (Barcelona) toward a comparative study of urban spaces, and then toward an interurban and eventually a global approach. He points out that while spatial metaphors like 'center' and 'periphery' continue to be seductive, their utility has been questioned by scholars of urban history in recent years. This point is particularly apposite for the late Habsburg Empire, where, as stated previously, cities like Cracow, Prague, or Budapest became functional equivalents of metropoles for their emerging national cultures.

The volume's remaining chapters consider Vienna itself in roughly chronological order, while integrating transnational dimensions in various ways. The chapters in Part Two analyze the creation of technical and scientific infrastructures in the metropolis from the 1850s to the 1870s from two complementary perspectives. In his chapter, Sándor Békési describes the beginnings of infrastructural expansion in Vienna following the failed revolution of 1848, particularly the major urban development projects leading up to and following the construction of the Ringstrasse, all of which, taken together, led to the creation of what he calls an "urban machine". As he shows, a transnational perspective is needed to understand this process, because technology transfers from multiple countries were essential to the completion of all of these projects. Békési argues further that a heterogeneous patchwork of overlapping knowledges from different disciplines was at work, blurring the boundaries between academic and administrative knowledge cultures. In her chapter, Marianne Klemun illustrates how such overlapping knowledges worked in practice. Focusing on the work of geologist and politician Eduard Suess, she shows that what she terms "metropolitan geology" played a dual role, providing basic research on Vienna's subterranean spaces and geological structures that proved useful for infrastructure projects such as the fresh water pipeline from the Alps or the development of sewer systems, and also supplying collections of "useful stones" displayed at the World Exhibition of 1873.

Part Three presents comparative studies of Vienna and other metropoles, as well as analyses of metropolitan networks and transnational knowledge circulation, beginning in the 1870s and continuing into the

1900s. In her chapter, Ulrike Spring discusses debates about the existence of an ice-free route through the Arctic Ocean in the late nineteenth century, focusing in particular on the Austro-Hungarian expeditions to the Arctic in the mid-1870s, led by officers Carl Weyprecht and Julius Payer, and the immensely popular public presentations of their results first in Hamburg and then in Vienna. In this chapter the concept of 'thick spaces' refers primarily to the public venues and the varied media such as maps and feuilleton articles in which this knowledge was presented, debated, and circulated in the metropolis. Rejecting any linear model of knowledge transmission between 'container'-like spaces, or from academic to public media, Spring suggests that it is more appropriate to speak of "entangled spaces of knowledge" in such contexts. In the following chapter, Petra Svatek takes a comparative approach, comparing the networks in the geographical sciences established in the universities of the two metropolitan capital cities Vienna and Berlin around 1900 with one another. As she shows, these two university institutes became centers of knowledge networks in geography, each of which operated transnationally by bringing in observations from within their respective nations and outside Europe, but they did not cooperate with one another to the extent that might have been expected, even after the transfer of Albrecht Penk from Vienna to Berlin in 1906. In her chapter, Brooke Penaloza-Patzak examines transnational *fin de siècle* networks established by Franz Boas and others in the United States, Berlin, and Vienna for the circulation and exchange of ethnographic artifacts, focusing particularly on objects made by peoples of the Northwest Coast of North America. She compares the ethnographic collections in the capital cities Berlin and Vienna, and shows how these and American metropolitan museums as well as the World Columbian Exhibition in Chicago (1893) became spaces of and media for the transfer of practical skills, methods, and theoretical approaches that came to define the new field of cultural anthropology.

Part Four returns the focus to Vienna, with studies of the circulation of knowledge among the metropolis' varied social spaces in the late nineteenth and early twentieth centuries. Johannes Mattes examines popular scientific societies, spatial transformations of the city and their impact on the development of knowledge-based public spheres. Focusing especially on speleology (cave research), a field that emerged at the margins of several sciences and the public sphere primarily in Vienna during the second half of the nineteenth century, and on two personalities who operated as go-betweens in this field, Adolf Schmidl and Franz Kraus, the chapter also investigates the function of spaces in-between for the communicative setting and the dissemination of scientific knowledge in metropolises. In the final chapter, Christian Stifter discusses the role of the sciences in the adult education movement in Vienna at the end of the nineteenth and the early twentieth centuries. As he shows, bourgeois liberals, among them leading academics, and social democrats formed a unique alliance in Vienna, bridging class boundaries by enlisting members of the educated middle classes and urban elites to establish popular education associations for working people, and recruiting aspiring younger scientists from the university as instructors. The crown jewel of these efforts was a new, well-equipped physical space for scientific adult education called the 'People's Home' (*Volksheim*) in the district of Ottakring. Teaching and learning there, Stifter argues, was based on a model of egalitarian cooperation taken from the French *Université Populaire* that compared favorably with classical secondary school and university instruction. In this analysis, the *Volksheim* was a "heterotope", an alternative social space in the sense defined by Michel Foucault, and thus in a certain sense a political project, despite the abstinence from party politics that was necessary to make it happen at all.

Conclusion

This volume presents new research on the establishment of infrastructures of scientific research and processes of knowledge transfer and exchange in the emerging imperial metropole Vienna in the nineteenth and early twentieth centuries. Taken together, these studies can be seen as examples of the co-evolution of the modern sciences and the leading city of the Habsburg Empire, especially in the dynamic years following the construction of the Ringstrasse. As suggested at the beginning of this introduction, the technical and scientific infrastructures described here can also be understood as preconditions for the explosion of creativity known as 'Vienna 1900'.

Of course one book cannot provide a comprehensive overview of this broad topic. Needed, for example, are integrative studies of links among research establishments and their access to funding and other resources, as well as more detailed comparative studies of other disciplines and their networks in this context. Particularly useful would be comparative studies of the empire-wide natural scientific survey projects mentioned previously and large-scale collaborative research in the humanities organized during the same period. Contrary to current 'two cultures' talk, the sciences and the humanities in this period appear to have shared a common interest in positivistic, documentation-oriented ways of knowing and therefore developed research practices that were more similar to one another than is currently realized. Suggestive in this regard is the use of terms such as 'archive', and even the word 'history' itself, in disciplines like botany, zoology, and geology during this period (Klemun 2015). Much needed, finally, is further work on the role of scientific, technological, and scholarly research in Vienna during the First World War (building on the first steps in Lange 2013; Soukup 2014; Mikoletzky and Ebner 2016).

It is to be hoped that this volume will stimulate further reflection and research on the co-creation of metropoles and the sciences and scholarship in the late nineteenth century more broadly, and also on the specific situation of Vienna.

Note

1. Ironically, as Johannes Mattes remarks (Chapter 9, this volume, endnote 1), the Academy of Sciences appears to have ceased using the sobriquet 'Imperial and Royal' in its own name and began referring to itself only as the 'Imperial Academy' (*kaiserliche Akademie*) as early as 1851, in the first issue of its official almanac (See Die Statuten der kaiserlichen Akademie der Wissenschaften 1851).

Bibliography

Aichner, C., and Mazohl, B. (eds.) (2017). *Die Thun-Hohensteinschen Universitätsreformen 1849–1860. Konzeption—Umsetzung—Nachwirkungen.* Vienna: Böhlau Verlag.

Ash, M. G. (1999). Die Wissenschaften in der Geschichte der Moderne (Antrittsvorlesung am Institut für Geschichte der Universität Wien, 2. April 1998). *Österreichische Zeitschrift für Geschichtswissenschaften* 10, 105–129. English abstract on p. 131.

Ash, M. G. (2000). Räume des Wissens: Was und wo sind sie? Einführung in das Thema. *Berichte zur Wissenschaftsgeschichte* 23, 235–242.

Ash, M. G. (2013). Jüdische Wissenschaftlerinnen und Wissenschaftler an der Universität Wien von der Monarchie bis nach 1945. Stand der Forschung und offene Fragen. In: Rathkolb, O. (ed.), *Der lange Schatten des Antisemitismus. Kritische Auseinandersetzungen mit der Geschichte der Universität Wien im 19. und 20. Jahrhundert.* Göttingen: V & R unipress, 87–116.

Ash, M. G. (2018). Multiple Modernisms in Concert: The sciences, Technology and Culture in Vienna around 1900. In: Bud, R., Greenhalgh, P., James, F., and Schiach, M. (eds.), *Being Modern: The Cultural Impact of Science in the Early Twentieth Century.* London: University College London Press, 23–39.

Ash, M. G., and Surman, J. (2012). The Nationalization of Scientific Knowledge in Nineteenth-Century Central Europe: An Introduction. In: Ash, M. G., and Surman, J. (eds.), *The Nationalization of Scientific Knowledge in the Habsburg Empire (1848–1918).* Basingstoke: Palgrave Macmillan, 1–29.

Beller, S. (ed.) (2001). *Rethinking Vienna 1900.* Oxford, NY: Berghahn Books.

Brantz, D., Disko, S., and Wagner-Kyora, G. (2012). Thick space: Approaches to Metropolitanism. In: idem. (eds.), *Thick Space: Approaches to Metropolitanism.* Bielefeld: Transcript, 9–27.

Brückweh, K., Schumann, D., Wetzell, R., and Ziemann, B. (2012). Introduction: The Scientization of the Social in Comparative Perspective. In: idem. (eds.), *Engineering Society: the Role of the Social and Human Sciences in Modern Societies, 1880–1980.* Basingstoke: Palgrave Macmillan, 1–40.

Buklijas, T. (2012). The Politics of *Fin-de-siècle* Anatomy. In: Ash, M. G., and Surman, J. (eds.), *The Nationalization of Scientific Knowledge in the Habsburg Empire (1848–1918).* Basingstoke: Palgrave Macmillan, 209–244.

Coen, D. R. (2007). *Vienna in the Age of Uncertainty: Science, Liberalism and Private Life.* Chicago: University of Chicago Press.

Coen, D. R. (2013). *The Earthquake Observers: Disaster Science from Lisbon to Richter*. Chicago: University of Chicago Press.

Coen, D. R. (2018). *Climate in Motion: Science, Empire, and the Problem of Scale*. Chicago: University of Chicago Press.

Cohen, I. B. (1987). *Revolution in Science*. Cambrdige, MA: Harvard University Press.

Czáky, M. (1996). *Ideologie der Operette und Wiener Moderne. Ein kulturhistorischer Essay zur österreichischen Identität*. Vienna, Cologne, and Weimar: Böhlau Verlag.

Dierig, S., Lachmund, J., and Mendelsohn, A. (2003). Introduction: Toward an Urban History of Science. In: idem. (eds.), *Science and the City (Osiris 18)*. Chicago: University of Chicago Press, 1–19.

Die Statuten der kaiserlichen Akademie der Wissenschaften. (1851). In: Kaiserliche Akademie der Wissenschaften (Hg.), *Almanach der kaiserlichen Akademie der Wissenschaften für das Jahr 1851*. Vienna: Braumüller, 1–11.

Feichtinger, J. (2010). *Wissenschaft als reflexives Projekt. Von Bolzano über Freud zu Kelsen: Österreichische Wissenschaftsgeschichte 1848–1938*. Bielefeld: Transkript.

Feichtinger, J. (2012). 'Staatsnation', 'Kulturnation', 'Nationalstaat': The Role of National Politics in the Advancement of Science and Scholarship in Austria from 1848 to 1938. In: Ash, M. G., and Surman, J. (eds.), *The Nationalization of Scientific Knowledge in the Habsburg Empire, 1848–1918*. Basingstoke: Palgrave Macmillan, 57–82.

Frank, T. (2018). Netzwerke zwischen Wien und Budapest. Die Medizinischen Fakultäten 1769–1945. In: Angetter, D., Nemec, B., Posch, H., Druml, C., and Weindling, P. (eds.), *Strukturen und Netzwerke. Medizin und Wissenschaft in Wien 1848–1955*. Göttingen: V & R unipress, 449–468.

Gantner, E. (2017). Logos, Industrial Palace, and Urania: The Urban Forms of Knowledge in Budapest. *Journal of Urban History* 43:4, 602–614.

Gantner, E., and Hein-Kircher, H. (2017). "Emerging Cities": Knowledge and Urbanization in Europe's Borderlands 1880–1945: Introduction. *Journal of Urban History* 43:4, 1–12.

Hammerl, C., Lenhardt, W., Steinacker, R., and Steinhauser, R. (eds.) (2001). *Die Zentralanstalt für Metereologie und Geodynamik 1851–2001. 150 Jahre Meteorologie und Geophysik in Österreich*. Graz: Leykam.

Hessler, M. (2010). "Damned Always to Alter, But Never to Be": Berlin's Culture of Change around 1900. In: Levin, M. R., Forgan, S., Hessler, M., Kargon, R. H., and Low, M. (eds.), *Urban Modernity: Cultural Innovation in the Second Industrial Revolution*. Cambridge, MA and London: MIT Press, 167–204.

Hochadel, O., and Nieto-Galán, A. (eds.) (2016). *Barcelona: An Urban History of Science and Modernity, 1888–1929*. London and New York: Routledge.

Hochadel, O., and Nieto-Galán, A. (eds.) (2018). *Urban Histories of Science: Making Knowledge in the City, 1820–1940*. New York and London: Routledge/Taylor & Francis.

Holzleitner, J. (2013). Die naturwissenschaftlichen Arbeiten des Kronprinzen Rudolf. Diploma thesis, University of Vienna. DOI: 10.25365/thesis25475.

Hughes, T. P. (1983). *Networks of Power: Electrification in Western Society 1880–1930*. Baltimore: Johns Hopkins University Press.

Jovanović-Kruspel, S., and Schumacher, A. (2017). *The Natural History Museum: Construction, Conception & Architecture*. Vienna: Naturhistorisches Museum.

Judson, P. M. (2016). *The Habsburg Empire: A New History*. Cambridge, MA: Harvard University Press.

Kandel, E. R. (2012). *The Age of Insight: The Quest to Understand the Unconscious in Art, Mind and Brain from Vienna 1900 to the Present*. New York: Random House.

Klemun, M. (2012). Introduction: 'Moved' Natural Objects—'Spaces in Between'. *Journal of the History of Science and Technology* 5, 9–16.

Klemun, M. (2015). Historismus/Historismen—Geschichtliches und Naturkundliches. Identität—Episteme—Praktiken. In: Ottner, C., Holzer, G., and Svatek, P. (eds.), *Wissenschaftliche Forschung in Österreich 1800–1900. Spezialisierung, Organisation, Praxis*. Göttingen: V & R unipress, 17–43.

Klemun, M. (ed.) (2020). *Wissenschaft als Kommunikation in der Metropole Wien. Die Tagebücher Franz von Hauers der Jahre 1860–1868*. Vienna, Cologne, Weimar: Böhlau.

Kos, W., and Gleis, R. (eds.) (2014). *Experiment Metropole—1873: Wien und die Weltausstellung*. Vienna: Czernin-Verlag.

Kraas, F. (2014). *Megacities: Our Global Urban Future*. Dordrecht (et al.): Springer.

Kraus, C. (ed.) (2004). *Der freie weite Horizont. Die Weltumseglung der Novara und Maximilians mexikanischer Traum*. Exhibition Catalogue. Bozen: Landesmuseum Schloss Tirol.

Lange, B. (2013). *Die Wiener Forschungen an Kriegsgefangenen 1915–1918*. Vienna: Verlag der Österreichischen Akademie der Wissenschaften.

Lefebvre, H. (1991). *The Production of Space*. Oxford: Blackwell.

Lenger, F. (2013). *Metropolen der Moderne. Eine europäische Stadtgeschichte seit 1850*. Munich: Beck.

Levin, M. R., Forgan, S., Hessler, M., Kargon, R. H., and Low, M. (2010). *Urban Modernity: Cultural Innovation in the Second Industrial Revolution*. Cambridge, MA and London: MIT Press.

Livingstone, D. N. (2003). *Putting Science in Its Place: Geographies of Scientific Knowledge*. Chicago: University of Chicago Press.

Livingstone, D. N., and Withers, C. W. J. (eds.) (2011). *Geographies of Nineteenth-Century Science*. Chicago: University of Chicago Press.

Mayntz, R., and Hughes, T. P. (eds.) (1988). *The Development of Large Technical Systems*. Frankfurt am Main and New York: Campus.

Mieg, H. A. (2010). Metropolen. In: Henckel, D., von Kuczowski, K., Lau, P., Pahl-Weber, E., and Stellmacher, F. (eds.), *Planen—Bauen—Umwelt. Ein Handbuch*. Wiesbaden: VS-Verlag, 322–325.

Mikoletzky, J. (1995). Vom Polytechnischen Institut zur Technischen Hochschule. Die Reform des technischen Studiums in Wien, 1850–1875. *Mitteilungen der Österreichischen Gesellschaft für Wissenschaftsgeschichte* 15, 79–100.

Mikoletzky, J. (2018). Von der Ingenieurschule zur Forschungsuniversität. Wandlungen der TH/TU Wien im 19. und 20. Jahrhundert. In: Feichtinger, J., Klemun, M., Surman, J., and Svatek, P. (eds.), *Wandlungen und Brüche. Wissenschaftsgeschichte als politische Geschichte*. Göttingen: V & R unipress, 111–118.

Mikoletzky, J., and Ebner, P. (2016). *Die Geschichte der Technischen Hochschule in Wien, 1914–1955. Teil 1. Verdeckter Aufschwung zwischen Krieg und Krise (1914–1937)*. Vienna: Böhlau Verlag.

Mühlberger, K. (2008). *Palace of Knowledge: A Historical Stroll Through the Main Building of the Alma Mater Windobonensis*. Vienna: V & R unipress.

Mühlberger, K. (2009). Das ‚Antlitz' der Philosophischen Fakultät in der zweiten Hälfte des 19. Jahrhunderts. Struktur und personelle Erneuerung. In: Seidl, J. (ed.), *Eduard Suess und die Entwicklung der Erdwissenschaften zwischen Biedermeier und Sukzession*. Göttingen: V & R unipress, 67–104.

Müller, G. B. (2017). Biologische Versuchsanstalt: An experiment in the experimental sciences. In: idem. (ed.), *Vivarium: Experimental, Quantitative and Theoretical Biology at Vienna's Biologische Versuchsanstalt*. Cambridge, MA: MIT Press, 3–18.

Nernes, R. (2010). Budapest. In: Gunzburger Makas, E., and Damljanovic Conley, T. (eds.), *Capital Cities in the Aftermath of Empires: Planning in Central and Southeastern Europe*. London and New York: Routledge/Taylor & Francis, 141–156.

Nyhart, L. K. (2009). *Modern Nature: The Rise of the Biological Perspective in Germany*. Chicago: University of Chicago Press.

Oldenziel, R. (2018). Whose Modernism, Whose Speed? Designing Mobility for the Future, 1880s–1945. In: Bud, R., Greenhalgh, P., James, F., and Schiach, M. (eds.), *Being Modern: The Cultural Impact of Science in the Early Twentieth Century*. London: University College London Press, 274–291.

Prokopovych, M. (2014). *In the Public Eye: The Budapest Opera House, the Audience, and the Press, 1884–1918*. Vienna: Böhlau Verlag.

Raj, K. (2016). Go-Betweens, Travellers and Cultural Translators. In: Lightman, B. (ed.), *A Companion to the History of Science*. Oxford: Oxford University Press, 39–57.

Ranzmaier, I. (2015). Die Philosophische Fakultät um 1900. In: Kniefacz, K., Nemeth, E., Posch, H., and Stadler, F. (eds.), *Universität—Forschung—Lehre* (650 Jahre Universität Wien, Vol. 1). Göttingen: V & R unipress, 133–148.

Reif, H. (2012). Metropolises: History, Concepts, Methodologies. In: Brantz, D., Disko, S., and Wagner-Kyora, G. (eds.), *Thick Space: Approaches to Metropolitanism*. Bielefeld: Transcript, 31–48.

Reiter, W. L. (2001). Stefan Meyer, Pioneer of Radioactivity. *Physics in Perspective* 3:1, 106–127.

Reiter, W. L. (2017). *Aufbruch und Zerstörung. Zur Geschichte der Naturwissenschaften in Österreich 1850 bis 1950*. Vienna: LIT.

Rentetzi, M. (2008). *Trafficking Materials and Gendered Experimental Practices: Radium Research in Early Twentieth-Century Vienna*. New York: Columbia University Press.

Rosner, R. W. (2004). *Chemie in Österreich 1740–1914. Lehre, Forschung und Industrie*. Vienna: Böhlau Verlag.

Rüdiger, J., and Schweizer, D. (eds.) (2015). *Stätten des Wissens. Der Weg der Universität Wien entlang ihrer Bauten 1365–2015*. Vienna: Böhlau Verlag.

Schivelbusch, W. (1988). *Disenchanted Night: The Industrialization of Light in the Nineteenth Century*. Berkeley: University of California Press.

Schorske, C. E. (1980). *Fin-de-Siécle Vienna: Politics and Culture*. New York: Alfred E. Knopf.

Schübl, E. (2010). *Mineralogie, Petrographie, Geologie und Paläontologie. Zur Institutionalisierung der Erdwissenschaften an österreichischen Universitäten, vornehmlich an jener in Wien, 1848–1938*. Graz: Leykam.

Seebacher, F. (2010). *Das Fremde im ‚deutschen' Tempel der Wissenschaften. Brüche in der Wissenschaftskultur der Medizinischen Fakultät der Universität Wien*. Vienna: Verlag der Österreichischen Akademie der Wissenschaften.

Sienell, S., Uhl, H., Taschwer, K., and Feichtinger, J. (2016). *Experimentalbiologie im Wiener Prater. Zur Geschichte der Biologischen Versuchsanstalt 1902–1945*. Vienna: Verlag der Österreichischen Akademie der Wissenschaften.

Soukup, R. W. (ed.) (2004). *Die wissenschaftliche Welt von gestern. Die Preisträger des Ignaz L. Lieben-Preises 1865–1937 und des Richard Lieben-Preises 1912–1928. Ein Kapitel österreichischer Wissenschaftsgeschichte in Kurzbiographien*. Vienna, Cologne, and Weimar: Böhlau Verlag.

Soukup, R. W. (2014). Das k. und k. Technische Militärkomitee im Spannungsfeld von Industrie und Wissenschaft. In: Matis, H., Mikeoletzky, J., and Reiter, W. (eds.), *Wirtschaft, Technik und das Militär 1914–1918. Österreich-Ungarn im Ersten Weltkrieg*. Vienna: LIT, 307–324.

Stadler, F. (ed.) (2019). *Ernst Mach: Life, Work and Influence*. Dordrecht: Springer.

Stráner, K. (2018). The Natural Sciences and Their Public at the Meetings of the Hungarian Association for the Advancement of Science in Budapest and Beyond, 1841–1896. In: Hochadel, O., and Nieto-Galán, A. (eds.), *Urban Histories of Science: Making Knowledge in the City 1820–1940*. Boca Raton, FL: Routledge/Taylor & Francis, 59–79.

Stratman, B. (2011). Megacities, Globalization, Metropolization, and Sustainability. *Journal of Developing Societies* 27:3–4, 229–259.

Štrbáňová, S. (2012). Patriotism, Nationalism and Internationalism in Czech Science: Chemists in the Czech National Revival. In: Ash, M. G. and Surman, J. (eds.), *The Nationalization of Scientific Knowledge in the Habsburg Empire (1848–1918)*. Basingstoke: Palgrave Macmillan, 138–156.

Surman, J. (2012). Science and Its Publics: Internationality and National Languages in Central Europe. In: Ash, M. G., and Surman, J. (eds.), *The Nationalization of Scientific Knowledge in the Habsburg Empire (1848–1918)*. Basingstoke: Palgrave Macmillan, 30–56.

Surman, J. (2019). *Universities in Imperial Austria 1848–1918: A Social History of a Multilingual Space*. W. Lafayette, IN: Purdue University Press.

Surman, J. (2020). Do Professors Walk? Academic Commuting and Urban Development in the Habsburg Metropolis 1848–1918. Paper presented at the Conference "Science in the Metropolis", Vienna 16 November 2017. Abstract and images available via www.academia.edu/42321641/. Download 25 March 2020.

Svobodný, P. (2018). Die akademische Migration zwischen den Medizinischen Fakultäten in Prag und Wien 1848–1945. In: Angetter, D., Nemec, B., Posch, H., Druml, C., and Weindling, P. (eds.), *Strukturen und Netzwerke. Medizin und Wissenschaft in Wien 1848–1955*. Göttingen: V & R unipress, 469–486.

Taschwer, K. (1997). Wie die Naturwissenschften populär wurden. Verbreitung naturwissenschaftlicher Kenntnisse in Österreich 1800 bis 1870. *Spurensuche. Zeitschrift für die Geschichte der Erwachsenenbildung und Wissenschaftspopularisierung* 8, 4–31.

Taschwer, K. (2019). *The Case of Paul Kammerer: The Most Controversial Biologist of His Time.* Schwartz, M. (trans.). Charlottetown and Montreal: Bunim & Bannigan.

Ther, P. (ed.) (2012). *Kulturpolitik und Theater. Die kontinentalen Imperien im Europa im Vergleich.* Munich: Oldenbourg and Vienna: Böhlau Verlag-Verlag.

Ursprung, J. (1984). Der 1. Internationale Ornithologen-Kongress 1884 in Wien. Ein Beitrag zur Geschichte der österreichischen Ornithologie. *Egretta* 27:1, 31–39. www.zobodat.at/pdf/EGRETTA_27_1_0031-0039.pdf. Download 23 March 2020.

Walch, S. (2016). *Triebe, Reize und Signale. Eugen Steinachs Physiologie der Sexualhormone. Vom biologischen Konzept zum Pharmapräparat, 1894–1938* (Wissenschaft, Macht und Kultur in der modernen Geschichte, Bd. 6). Vienna: Böhlau Verlag.

Wiener Zeitung (17 May 1847). Nr. 136, 1085–1086 (Patent of the Imperial and Royal Academy of Sciences).

Wise, M. N. (2018). *Aesthetics, Industry and Science: Hermann von Helmholtz and the Berlin Physical Society.* Chicago: University of Chicago Press.

Withers, C. W. J., and Livingstone, D. N. (2011). Thinking Geographically about Nineteenth-Century Science. In: Livingstone, D. N., and Withers, C. W. J. (eds.), *Geographies of Nineteenth-Century Science.* Chicago: University of Chicago Press, 1–19.

Part One

Historiographical Overviews

2 Metropolitan Natural Histories

Inventing Science, Building Cities, and Displaying the World

Dorothee Brantz

Introduction

In a recent book on the writing of global history, Lynn Hunt evocatively traced how the discipline of history has evolved from its traditional nineteenth-century obligation to narrate—and by extension educate—the nation to current concerns about diversity in an age where the highly individualized self has to be constantly (re)negotiated vis-à-vis an increasingly global network of identifications (Hunt 2014). History as a discipline fulfills an important role in this continuous (re)negotiation; and the history of science, I would argue, holds a particularly central position in this debate because of its critical focus on knowledge accumulation about the changing scientific understanding about humans, nature, and society. If one wants to look at the concrete geographical and material circumstances of this kind of scientific knowledge production, cities are a terrific place to start. Cities, particularly the aspiring metropoles of the nineteenth century, have been key arenas where such debates about the intersection of the human and natural worlds were carried out in the newly emerging institutional contexts of universities, museums, and government agencies.

Natural history offers a particularly rich field of investigation because it provides telling insights into the shifting interpretation of human-nature relations and how they were embedded in distinct networks of power, architectural representation, scientific discoveries, and political institutionalization, all of which manifested themselves most poignantly in cities. In the following chapter I would like, first, to look at the meaning of the terms 'metropolis' and 'metropolitanism' as historical phenomena within expanding network societies. Second, I will turn to natural history museums and zoos as specific empirical sites for the representation of metropolitan science. I will end by proposing the notion of "thick spaces" (Brantz et al. 2012) as a conceptual and methodological tool to articulate the intricate relationships between science and metropolitan networks.

1. The Metropolis as Historical Phenomenon

The origin of the term 'metropolis' reaches back to Greek antiquity. Referring to the 'mother city' of a colony, the metropolis denoted the place where new settlers had originally come from; hence it created a link between places of origin and destination within an imperial context (Farías and Stemmler 2012). In addition to this ancient imperial connotation, in the early modern period there was also a religious identification, because metropolis meant the seat of the bishop. Much more common than these ancient imperial and early modern religious identifications, however, is its modern meaning, which equates the metropolis with a political, economic, and cultural center—the seat of government, key societal institutions, and as the dominant place of commerce, trade, and economic production. The modern metropolis is also closely associated with the idea of the nation. Nineteenth- and twentieth-century metropoles usually functioned as the central city in a nation, often its capital city. They served as a kind of magnet that drew people, ideas, and resources from elsewhere to the center. In that sense, modern metropoles reversed the ancient notion of the 'sending out' function of the mother city. Indeed, the question of push and pull factors is very pertinent to our contemporary understanding of the meaning of 'metropolitan', because these push and pull factors lead to the complex entanglements that linked cities to other cities and regions across the globe.

As a modern concept, metropolitanism is also closely associated with large-scale processes of transformation—namely urbanization, mobilization, industrialization, imperialism, capitalism, and of course also the growing scientization (*Verwissenschaftlichung*) of society (Bud et al. 2018; Frisby 2001). This scientization entailed not only the growing interest in scientific investigations but also the expanding organization of urban life according to scientific principles, for instance through the implementation of public hygiene regimes or the rationalization of work processes through Taylorism. As such metropolitanism is linked to another rather amorphous concept—'modernity' with its close associates 'modern' and 'modernism'. As scholars ranging from the Frankfurt School to Michel Foucault and Bruno Latour have argued, the modern project gave rise to a contradictory web of power relations that artificially separated nature and culture to establish a hegemonic system of instrumental reason and politicized order (Adorno and Horkheimer 1944; Foucault 1970; Latour 1991). The rising metropoles of the nineteenth century became central sites where this modern project was facilitated, not the least through the growing impact of science and technology (Frisby 2001; Dierig et al. 2003; Levin et al. 2010). As Lynn Nyhart has argued, the nineteenth century witnessed a close relationship between the emergence of modern society and the construction of the modern vision of nature, particularly but not only in Germany (Nyhart 2009).

If we start with the premise that metropoles are large urban agglomerations with populations of several hundred thousand inhabitants, then London, Paris, St. Petersburg, Vienna—all capital cities—were the metropoles of the nineteenth century. By 1900, a few others had joined their ranks. Berlin—also a capital city—New York, and Chicago had moved up to the top five of the most populated cities in the world, which underscored the increasing centrality of North America when it came to rapid metropolitan developments. But population size was of course only one factor in the determination of a metropolis, which was illustrated, among other things, by the continuous debate about whether Berlin and Chicago could even be regarded as metropoles at all (Brunn and Reulecke 1992; Cronon 1991; Large 2000; Miller 1996; Pacyga 2009).

According to historian Heinz Reif, metropoles are characterized by a "surplus" of numerous factors. Reif argues that: "metropolises are sites of structural wealth in material and cultural resources," which includes infrastructures and cultural institutions, even more so since "a metropolis is a site of the highest national and international political, economic and, above all, (high) cultural centrality (of elites)" (Reif 2012, 32–33). In other words, metropolises are centers of multiple interlinked productions. Often, they are the seats of government and other important institutions including universities, research institutes, and national museums. The combination of all of these institutions promotes a high density and diversity in the flow of intellectual activities and the proliferation of expert knowledges. Given their institutionalization in such locations, these knowledge productions often play into hegemonic discourses of cultural self-reproduction and national legitimization rather than promoting a critically reflective perspective on society. Put differently, these institutions endorsed the metropolitan spirit to which they had given rise in the first place.

We should, however, be careful not to understand the metropolis, its museums and zoos as 'containers' where living and dead animals were collected, concentrated, produced, and put on display. For a long time, scholars have focused on the internal dynamics, problems, and contradictions that drove metropolitan developments. The idea that the metropole functioned as a "laboratory of modernity" underscores this standpoint because it alludes to the demarcated space of the laboratory where experimental activities take place under expert supervision, even if not necessarily with predictable outcomes (Matejovski 2000; Schrock 2019). This perspective has generated tremendous insights into the internal operations of cities. In recent years, though, this internal focus is being augmented by another perspective that looks the opposite way to examine the external (particularly imperial) connections that made this metropolitan growth of western cities possible in the first place (Driver and Gilbert 1999; Schneer 1999). This perspectival shift goes hand in hand with a more globalized understanding of the world (Bayly 2004; Osterhammel 2015).

Consequently, urban agglomerations—particularly metropoles—are increasingly analyzed in terms of *relationships and interactions*, be they political, economic, socio-cultural, or environmental (see also the chapter by Oliver Hochadel, Chapter 3 in this volume). Urban scholars have long emphasized the relational character that undergirds urban dynamics, but interestingly, the impact of ecological factors has been neglected (to put it mildly) until recently. Conventionally, the city has been viewed as a built environment driven by human activity. In the past decade, a growing number of scholars are insisting that cites are places where nature is not only staged, but where it is actually invented through the hybridization of multiple (cultural, economic, environmental, political, religious, and social) factors (Brantz 2007; Gandy 2003; Heynen et al. 2006).

This broadened perspective opens up new opportunities for rethinking the role of science in the city and the metropolis, which is one of the central aims of this book. Let us proceed with the following three premises: first, urban dynamics grow out of relations, including socio-natural hybridizations. In other words, we should not understand metropoles and their institutions as geographically demarcated 'containers', but as networks of relations. Second, when we speak of 'density' and 'diversity'—two notions that have always been considered primary metropolitan characteristics— we should also incorporate the non-human, most notably animals and plants. The notion of 'diversity', which has always been so highly valued among progressive urbanites and scholars, becomes even more open and inclusive when we do this. Third, the metropolitan is constituted through a complex web of entangled local-regional-international and global networks that generated the dynamic upon which metropolitan cultures are built. Metropoles are not just geographical places, but—and perhaps more importantly—they should be understood as a tapestry of globally expanding (imperial) networks. They functioned as hubs or nodes in a network. In the following section I will turn to natural history museums and zoos to show specifically how they contributed to these networks on numerous levels.

2.   Natural History Museums and Zoos as Nodes in Networks

In the course of the nineteenth century, natural history museums and zoos were founded all over Europe and later also in North America and other continents. Among the first natural history museums were those in Paris, London, Berlin, and later Vienna, New York, and Chicago (Asma 2001; Jovanović-Kruspel and Schumacher 2017; Riedl-Dorn 1998; Yanni 1999). The first zoo, in London's Regent's Park, was founded in 1828. (The zoological garden in Paris, often called the first modern zoo, was established alongside the *Jardin des Plantes* there in 1793.) Other cities soon followed and throughout the century, zoological gardens sprang up in most aspiring metropoles (Akerberg 2001; Ash 2008, 2018; Baratay 2002; Hanson 2002; Miller 2013; Rosenthal 2003; Rosenthal 2003; Wessely 2008). Natural history museums and zoos served as telling examples of how such a metropolitan spirit was generated in the course of the nineteenth and early twentieth century. Much has been written about each of these establishments, but rarely have they been viewed as related institutions, even though they have much in common. Exceptions may be found in a few edited collections, but even there, most of the essays discuss one institution or the other, but not both in comparison or in relation with one another. One of the common features of these institutions lies in the way they represent nature through particular animal-centered displays (Macdonald 1998; Thorsen et al. 2013). Natural history museums and zoos attested to the growing concentration of (natural) resources in cultural/scientific institutions that were established for elite and increasingly also mass audiences. Indeed, the foundation of natural history museums, and even more so, of zoos was closely linked to the metropolitan aspirations of the urban elites who founded them, especially in capital cities where these institutions became the central sites of collection.

Many of the modern natural history museums and zoos that were founded all across the globe during the nineteenth century were created as institutions for knowledge accumulation and the subsequent dissemination of such knowledges to research communities and urban publics. Indeed, they attested to the expanding notion of an urban public that comprised not only bourgeois men and women, but also children as well as the lower social classes. By learning about the natural world, they were also supposed to be educated about bourgeois morals and values. Combining science, education, and entertainment, zoos and natural history museums offered entertaining training grounds for the bourgeoisification (*Verbürgerlichung*) of urban society. They enhanced the presence and impact of nature in the city because they functioned as actual places of (second) nature in the urban realm. Zoos in particular became popular sites for the supposed experience of the natural flora and fauna of one's home and foreign lands. For that reason, many zoological displays were placed in sprawling parkscapes where the public could promenade, frolic, and wander. Obviously, these institutions mainly promulgated imaginary visions of nature that were tamed, cleaned up, and professionally staged. But most zoos also emphasized their educational aspirations. School children but also the public at large were supposed to learn about different animals, their behavior and their places of origin. In natural history museums the educational purpose was perhaps more straightforward, because their expositions were not so veiled in entertaining surroundings as zoos, where music pavilions, restaurants, and cafés also offered lively attractions (Wessely 2008).

In addition to their educational function, both of these institutions served research scientists. Large parts of many natural history museums were inaccessible to the public and devoted to scientific research. Indeed, this architectural separation of public and scientific wings within natural history museums was not without controversy, as the example of Berlin underscored (Kretschmann 2006). One of the principal missions of the zoology departments of natural history museums was the development and augmentation of classificatory systems as well as the investigation of animal anatomy. Closely related to this study of dead animal bodies was the practice of taxidermy as a measure to preserve specimens (Poliquin 2012).

Within natural history museums, the layout of the buildings indicated how knowledge about the natural world was to be acquired. For scientists this was to be done through laboratory-like settings with microscopes and other investigative tools set up in a separate environment of artifacts stored in wooden closets that provided their distinct systematized arrangement to the collections. In the public sections of the museum, the layout of the rooms and the particular staging of animal displays in glass vitrines also suggested a distinct order of significance, an order that placed large mammals at the center. Rather than encountering rooms overloaded with everything the museum had amassed, visitors were presented with a selection of exhibition pieces that were systematically arranged to present the world in a comprehensible order that was supposed to make the (natural) world readable (Bennett 1998). Both the scientific and public collections emphasized reading as the primary mode of orientation and knowledge building. Visitors were supposed to be able to read the collection like an open book.

Nineteenth-century zoos also fulfilled specific scientific functions because they offered an opportunity to study the physiology and behavior of living animals for the first time. Countless biologists, animal behaviorists, physiologists, ornithologists, veterinarians went to zoos to study animals, especially those from places far away. To many newly emerging sciences, museums and zoos offered staged spaces of knowledge production and distribution, and hence they also contributed to the scientization (*Verwissenschaftlichung*) of society (Daum 1998; Nyhart 2009). Presenting projections of nature through living creatures and dead specimens that had been extracted from the wild across the globe, both of these institutions exemplified how the world was brought to and narrated in the city. These projections of nature also served to 'naturalize' imperial visions independently of actual geopolitics (Palló 2012).

This narration was tied into several interconnected networks that linked metropolitan centers to the rest of the world. In a well-known 2003 collection of essays on *Science and the City*, the volume's editors Sven Dierig, Jens Lachmund, and Andrew Mendelsohn laid the foundation for a more integrative investigation of the historical entanglements between science and urban development. Arguing for a perspective centered on the "coproduction" of science and the city, this collection of essays presented numerous perspectives on how scientific discourses and practices

contributed to the production of urban life. In conceptualizing this coproduction, they emphasized the need to pay close attention to the local as
a central point of investigation because the local offers a place-specific
arena for investigations of the interchanges between science and the city.
With regard to the local, place and location are crucial aspects, both
in relation to the physical parameters of a scientific locale (observatory,
laboratory, museum, zoo etc.) and their position in the urban landscape
including their infrastructural integration. Natural history museums were
frequently built in prime locations that signified their central position in
a city's museum landscape. For instance, New York's American Museum
of Natural History is prominently located on Central Park West. Vienna's
Naturhistorisches Museum is part of the Ringstrasse ensemble, located
across from the Hofgarten and directly opposite the city's famous art
history museum (Jovanović-Kruspel and Schumacher 2017). In Washington, the National Museum of Natural History is part of the Smithsonian
Institution located on the National Mall. In London, the natural history
museum is also part of a larger complex of museums in South Kensington
(Thackray 2013). Obviously, displays of natural history were considered
significant for a city's urban identity and representation. In some cities,
such as Paris's *Jardin des Plantes*, natural history museums and zoos were
even placed together to encourage comparative observations by scientists
and the public (Spary 2000). Nineteenth-century metropolitan zoos were
usually not as centrally located, simply because they required more space,
but they often also occupied prominent locales like New York's Central
Park, London's Regents Park, Vienna's Schönbrunn castle park, or Chicago's Lincoln Park. In Berlin, the zoo was initially 'jwd', an expression
for really far away (*janz weit draußen*), but by the end of the century it
was at the heart of the city's new western center around the Zoologische
Garten train station, a central traffic hub of the city (Bruce 2017).

What does all this have to do with the history of science, one might
ask. For one, it underscores the representational significance of science in
urban spaces. Museums played a crucial role in the reinvention of cities
as metropolitan centers, which was demonstrated, among other things,
by their location. In addition to location, architecture served as another
marker of significance. Many of the natural history museums that were
created from the mid-nineteenth to the early twentieth century were built
in a neoclassical style that underscored their representational function
as 'temples of science'. Looking at the examples of Vienna, Berlin, New
York, London, and Chicago, one is struck by their similarly neoclassicist
architecture, with impressive entrance portals, columns, and richly decorated façades.

To illustrate the significance of architecture for the urban representation
of science, let us look briefly at a specific example, Vienna's *Naturhistorisches Hofmuseum* (later Imperial and Royal Natural History Museum)
(Figure 2.1).

Figure 2.1 Imperial and Royal Natural History Museum, Vienna, 1901.
Source: Botanik und Zoologie in Österreich 1850 bis 1900 (Festschrift). Vienna 1901. Public domain.

As stated previously, the natural history museum was part of the Ringstrasse ensemble that was built as part of the city's extensive urban modernization during the nineteenth century. This new museum complex, to which the *Naturhistorisches Hofmuseum* as well as the *Kunsthistorisches Museum* belonged, was a highly prestigious building project originally designed by Gottfried Semper, who together with Karl Freiherr von Hasenauer had envisioned it as a representational imperial triangle. Opened with great pomp by Kaiser Franz Josef in 1889, it remained one of the Imperial and Royal monarchy's key sites to represent the cultural and scientific prowess of the Habsburg Empire until its end in 1918 (Jovanović-Kruspel and Schumacher 2017). A unique feature of the Vienna natural history museum was its elaborate façade that encouraged a closer engagement between the building's architecture and its interior design and exhibitions. Semper and Hasenauer had decided to divide the façade into distinct levels that would symbolize different aspects of scientific discoveries and innovations. Depending from which side one approached the building, one was presented with a different chronological perspective on the provenance of science. As Iris-Amelie Ginthör-Weinwurm (2008) has shown, the diverse sculptures on the four outside façades of the building document the contemporary understanding of scientific knowledge production as a consecutive and progressive endeavor that led from the ancient to the modern period. With their depiction of scientific icons, these façades represented the underlying ideology of late nineteenth-century science that was premised on an authoritative truth driven by progressive notions of knowledge accumulation by great (white) men who build on their predecessors' knowledge and who worked in close conjunction with important political leaders, in Vienna's case absolutist monarchs (Supuković 2015, 64). These sculptured façades created an authoritative narrative about the presence of a distinct vision of science in urban space—one that still dominates today, in large part because of this architecture. These kinds of museums are among the most powerful and lasting (architectural) embodiments of scientific narratives in urban space.

The interior extended this discourse on the natural world through its displays of zoological specimens, plants, and minerals. As scholars have shown, the staging of animal displays documented bourgeois urban notions of nature (Köstering 2003; Kretschmann 2006; Supuković 2015). Animals were arranged in elaborate dioramas to guide the visual imagination of visitors and to depict natural history as a continuous evolutionary development through the objectification of dead animal bodies. Local animals like deer were frequently displayed as nuclear families in natural settings to underscore notions of home.

Another example of the architectural staging of animals in cities was the elephant pagoda in Berlin's zoological garden (Figure 2.2). Built in 1873, it exemplified a particular exoticist aesthetic preference in zoo architecture

Figure 2.2 Berlin Zoological Garden, Elephant House, 1875.

Source: Public domain. https://commons.wikimedia.org/wiki/Category:Elefantenpagode_%28Zoologischer_Garten_Berlin%29#/media/File:Zoo_Berlin_Elefantenhaus_um_1875. Download: November 2, 2019.

at the time. This architectural style amalgamated the functional necessities of housing animals with framing them in a specific cultural context related to their alleged homelands. Since elephants often originated in India, housing them in a pagoda—an Indian temple—appeared like a suitable setting.

The intended purpose was to teach the visiting public not only something about the animals on display but also about the culture of their supposed homelands. In Berlin, not only elephants but many other animals were housed in such structures. Giraffes and antelopes were housed in a Moorish temple, bison in Native American–styled huts, chicken and pigs in German countryside cottages (*Fachwerkhäuser*) (Klös et al. 1990; Wessely 2008). This conflation of animal displays and cultural signification underscored how zoos functioned to cultivate nature and simultaneously naturalize culture.

In general, the architecture of these institutions was supposed to readily suggest their higher purpose to the visiting public. The—often imposing—architecture of natural history museums and zoos was to imbue visitors with a sense of awe not only about the grandeur of nature, but also about the greatness of those who had put these collections together. These institutions were supposed to represent humans' power over nature and underline the power of (scientific) knowledge to its (bourgeois) public. Natural history museums and zoos were particularly telling places in this regard, because they combined, each in its own way, knowledge production with a sense of wonder. For those who caught on to such a sense of wonder, these places opened up fascinating spaces for discovery and inspiration. For others, the natural history museum or the zoo could be a place of information or just for simple entertainment. Both were aimed to present the natural world according to an understanding of urban social order within globalizing networks. As the façade of Vienna's natural history museum and Berlin's elephant pagoda illustrate, these kinds of scientific and architectural representations pointed well beyond their cities—they hinted at the inter- and transnational connections upon which these facilities were built. This was illustrated even more impressively by the animals that had been brought to these urban centers from every corner of the globe. Zebras from the African savannah, grizzly bears from the mountains of North America, koalas from Australia, or even penguins from the South Pole were brought to zoos as living creatures or to natural history museums as dead specimens.

Given the special popularity of exotic animals, their acquisition played a key role in the growing business of animal trading. By the end of the nineteenth century, the trade of animals and specimens developed into a global network. Particularly the German animal trader and zoo founder Carl Hagenbeck had created a sizable international animal trade empire spanning the globe (Ames 2008, Dietrich and Rieke-Müller 1998). Hagenbeck, who had taken over his father's business, soon started to take advantage of the expanding global networks that connected his native Hamburg with many other parts of the world. Sending out animal hunting expeditions to Africa and later many other areas of the world, Hagenbeck became the most renowned animal trader of his generation, supplying zoos and other animal attractions in Germany and later all over the world. Hagenbeck himself was a businessman rather than scientist, but his expeditions frequently included scientists and scientific institutions from cities across Europe and North America. Most living animals went to zoos and menageries. Many of those who had died in transit went to natural history museums. Hagenbeck's main interest centered on profitable entertainment, but in the course of his career he also revolutionized the staging of zoo animals in his newly founded "animal park" in Hamburg Stellingen, where animals were kept in enclosures without bars (Rothfels 2002). Hagenbeck might not have been a classical scientist, but his activities certainly helped to bridge the worlds of science and entertainment. Apparently, his contemporaries valued his contribution to the sciences; in 1912, he was admitted into the prestigious German Academy of Natural Scientists Leopoldina.

Hagenbeck's trading activities underscored the growing globalization of animal trading. An animal might have been captured in Africa, then brought to a European zoo, and after a couple of years sold to another zoo or even to a circus in the Americas, thus becoming part of an international network of trade relations. One of the most famous examples of such an international transfer was Jumbo, the elephant. Born in Sudan in 1860, he was brought to the *Jardin des Plantes* in Paris at age four. Just a year later, in 1865, he was taken to London's Regent Park Zoo, where he became a major public attraction. When he was sold to the American entrepreneur and circus owner Phineas T. Barnum of Barnum & Bailey Circus in 1882, thousands of Londoners protested that their favorite zoo animal had been sold across the Atlantic (Chambers 2008; Harding 2000). In fact, Jumbo's story illustrates not only international trade, but also the global dimension of these kinds of animal exchanges because he was involved in a far more elaborate network of exchanges than his geographic locations might suggest. The hunters and traders who had killed his mother, captured him, and transported him across continents came from Sudan, Italy, Germany, France, Great Britain, the US, and Canada. His keepers were British, Indian, and American. While Jumbo's fate might have been exceptional, his story was not. Millions of animals were captured all over the world, shipped to far off locations for scientific and exhibition purposes and if they died on the way, which countless numbers of them did, then they could still serve as scientific specimens. Many were also hunted and killed right away to be delivered to metropolitan institutions, or sometimes even on order from them. The capture, trade, and sale of animals developed into a profitable (even if risky) industry in the course of the nineteenth century. Through these growing global networks of trade, zoos and natural history museums became part of a system that reconfigured the world according to the expanding hegemonic aspirations of rising western metropoles.

Obviously, imperialism played a crucial role in these hegemonic pursuits, and hence in the development of western natural history museums and zoos. Indeed, one might argue that without the expansive ideologies of imperialism, zoos and natural history museums would not have emerged at this particular moment. Colonial expansion opened new areas for travel for westerners, and it also led to the massive construction of transport infrastructures, such as railways, which made it possible to bring animal specimens to European and North American cities in the first place. The zoos and natural history museums that became the new homes of these specimens helped to bring the (natural) *world TO the city*, and as such they became deeply implicated in these imperial fantasies and in the violence they fostered. The millions of animals who were imported from the countryside and foreign lands to the centers of western metropoles like Berlin, London, New York, Paris, and Vienna added diversity to the city, either as real creatures or as dead specimens, but they also embodied, quite literally, the enslavement of the natural world, foreign lands, animals, and peoples. The then highly popular people shows (*Völkerschauen*) were yet further embodiments of this exploitation in the name of science and entertainment (Dreesbach 2005; Schwarz 2001).

These institutes also represented the *world IN the city*. Many zoo advertisements claimed that one could take a trip around the world within an afternoon by visiting the animals and gardens on display. Moreover, zoos and natural history museums participated in the imaginary mapping of the world by displaying animals according to their geographic region of origin. Put differently, they created miniature worlds mapped onto the landscape of the city. The emerging science of animal geography and ecology further attested to shifting notions of zoology and their institutionalization in growing international networks that increasingly promoted the exchange of living and dead animal bodies as well as the exchange of ideas and knowledge among scholars.

For instance, Berlin, like most cities with major zoos and natural history museums, was not just any city, but a capital city at the heart of a nation. This connection was not coincidental. Given that capital cities are not simply centers of government, but also the central sites of national representation, museums and zoos were significant for this representation of national identity. According to Rudolf Virchow—professor of anatomy, Reichstag representative, and one of Germany's most eminent social reformers of the time—the public sections of the natural history museum's collection should serve as educational spaces and places for the national representation of science (Köstering 2003, 50). In 1888, a government decree ordered that all specimens collected during state-sponsored expeditions had to be turned over to the museum. The aim was to establish the Berlin museum as THE central representational site for natural history in Germany. The same was true in most other capital cities. This underscored the cultural significance of scientific institutions and it helped to construct the metropole as the heart of the nation in relation to the rest of the nation as well as in competition of nations and empires. As such they, much like zoos, were not just to represent the world, but also the homeland (*Heimat*) by transgressing the boundaries between inside/outside and self/other. They were supposed to teach urban dwellers about the nature of their nation. For the zoo this education about "German nature" was part of the explicit mission. Deer, foxes, boar, and other creatures of the woods were placed side by side with chickens, horses, goats, and cattle to represent the richness of the domestic German fauna. The natural history museum followed a similar approach to presenting domestic and foreign animals, even though in the museum zoological categorizations of class, order, family, and species were more important than geopolitical classifications. Dioramas were a broadly favored method of displaying animals. The widespread popularity of dioramas around the turn of the twentieth century highlighted the growing emphasis on visual displays and the gradual shift from 'reading' descriptions to seeing proscribed connections. In these dioramas, animals were usually staged as family groups and placed in painted settings that were supposed to depict the animals' natural habitats (Metzler 2007; Mothes 2013; Quinn 2006; Scheersoi and Tunnicliffe 2019).

Zoos were also about looking, but in them hearing, to some extent smelling or even touching, were also significant aspects of the experience. This more holistic sense perception contributed to the entertaining rather than straightforwardly scientific character of zoos. Much like in museums, the educational dimension was emphasized through the labels that provided information about the common names of the animals and their zoological classification as well as their area of origin. To some extent, science also took place behind the scenes of the public displays. It centered on measuring, classifying, and at the end of the century also increasingly breeding animals in captivity, but as mentioned previously, it also entailed the study of animal behavior. Overall, whereas natural history museums might have pursued a more explicit scientific agenda, zoos, too, participated in the production of scientific knowledge about animals. Undoubtedly, most of the knowledge that was generated in these institutionalized settings was driven by strictly human, often explicitly anthropomorphic, visions of nature and human hegemony. As such it always also reflected the particular epistemologies about nature that were prevalent at the time, and they do so to the present day, when notions of nature conservation, climate change, and species extinction are increasingly taking center stage and citizen science offers a new means for public engagement and scientific debate.

Conclusion—'Thick Spaces'

Natural history museums and zoos were nodes of complex networks comprising a number of levels from the very local all the way to the global. Scientifically, each of these institutions was incorporated in distinct inner-urban networks of scientific knowledge production and dissemination, regional ties as well as international networks of scientists and imperial animal/specimen exchange. Such animal displays depended on very local institutional and architectural frameworks as well as on their regional, international, and even global interrelationships, which underscored the complexity of the metropolitan entanglements that were usually premised on hegemonic claims and hierarchical inequalities.

In closing, let us briefly consider the notion of 'thick space', which might offer both a conceptual frame and methodological tool to capture the complexity of such urban network relations, particularly in relation to metropolitan histories. Clifford Geertz identified thick descriptions as an ethnographic method to unearth a web of meaning in human behaviors through analyzing human activities in their cultural context. What exactly can that mean in an urban context? Of course, it can entail a close description and subsequent analysis of specific urban practices. In that sense a Viennese Urania program, the activities of the Vienna popular education center called the "people's home" (*Volksheim*) in the working-class district of Ottakring (Stifter, Chapter 10 in this volume), or the linkages of urban geological analyses and the construction of Vienna's alpine water viaduct (Klemun, Chapter 5 in this volume) could be treated much like Geertz's close studies of cultural practices like a Balinese cockfight (Geertz 1973). Apart from investigating individual human behaviors and their socio-cultural contexts as a classical thick description would pursue, an urban perspective focusing on a notion of "thick space" (Brantz et al. 2012) might also center on a detailed analysis of networks and their relationships across different scales of engagement ranging from the local to the global (Spring, Penaloza-Patzak, and Mattes, Chapters 6, 8, and 9 in this volume). Such thick descriptions premised on a multiplicity of urban scales focus on the spatial dimensions—how spaces are constructed, debated, represented, and lived. And such a focus centers not only on the material dimensions of space but also on spatial relations. After all, as Henri Lefebvre insisted: "(social) space is a (social) product" (Lefebvre 1991, 26). As urban scholars we need to ask how these productions take place; under which circumstances; who the key actors were; and how the meaning of these produced spaces changed over time. This kind of analysis warrants a thick description that pays attention to many of the historical intricacies and contingencies as well as developments across different scales. Geertz insisted that a key aspect of a thick description is the inherent interpretation of culture that has to take place. As many chapters in this volume demonstrate, such an interpretation of culture is also pertinent for the understanding of metropolitan histories and scientific entanglements. Metropolitan scientific discourses and practices might offer a particularly telling arena to examine such entanglements across scales and disciplinary boundaries.

Bibliography

Adorno, T., and Horkheimer, M. (1944). *Dialectic of Enlightenment*. New York: Continuum.

Akerberg, S. (2001). *Knowledge and Pleasure at Regent's Park: The Gardens of the Zoological Society of London during the Nineteenth Century*. Umea: Umea University Press.

Ames, E. (2008). *Carl Hagenbeck's Empire of Entertainments*. Seattle: University of Washington Press.

Ash, M. G. (ed.) (2008). *Mensch, Tier und Zoo: Der Tiergarten Schönbrunn im internationalen Vergleich vom 18. Jahrhundert bis heute*. Vienna: Böhlau.

Ash, M. G. (2018). Zoological Gardens. In: Curry, H. A., Jardine, N., Secord, J. A., and Spary, E. C. (eds.), *Worlds of Natural History*. Cambridge: Cambridge University Press, 418–432.

Ash, M. G., and Surman, J. (eds.) (2012). *The Nationalization of Scientific Knowledge in the Habsburg Empire, 1848–1918*. Basingstoke: Palgrave Macmillan.

Asma, S. (2001). *Stuffed Animals and Pickled Heads: The Culture and Evolution of Natural History Museums*. New York: Oxford University Press.

Baratay, E. (2002). *Zoo: A History of Zoological Gardens in the West*. London: Reaktion Books.

Bayly, C. (2004). *The Birth of the Modern World, 1780–1914: Global Connections and Comparisons*. Oxford: Oxford University Press.

Bennett, T. (1998). Speaking to the Eyes: Museums, Legibility, and the Social Order. In: Macdonald, S. (ed.), *The Politics of Display: Museums, Science, Culture*. London: Routledge, 25–35.

Brantz, D. (2007). The Natural Space of Modernity: A Transatlantic Perspective on (Urban) Environmental History. In: Lehmkuhl, U., and Wellenreuther, H. (eds.), *Historians and Nature: Comparative Approaches to Environmental History*. Oxford: Berg Publishers, 195–225.

Brantz, D., Disko, S., and Wagner-Kyora, G. (eds.) (2012). *Thick Space: Approaches to Metropolitanism*. Bielefeld: Transcript.

Bruce, G. (2017). *Through the Lion's Gate: A History of the Berlin Zoo*. New York: Oxford University Press.

Brunn, G., and Reulecke, J. (1992). *Metropolis Berlin: Berlin als deutsche Hauptstadt im Vergleich europäischer Hauptstädte 1871–1939*. Bonn: Bouvier.

Bud, R., Greenhalgh, P., James, F., and Shiach, M. (eds.) (2018). *Being Modern: The Cultural Impact of Science in the Early Twentieth Century*. London: UCL Press.

Chambers, P. (2008). *Jumbo: The Greatest Elephant in the World*. Westminster: Steerforth.

Cronon, W. (1991). *Nature's Metropolis: Chicago and the Great West*. New York: W.W. Norton.

Damaschun, F. (2010). *Klasse, Ordnung, Art: 200 Jahre Museum für Naturkunde.* Rangsdorf: Basilisken Presse.

Daum, A. (1998). *Wissenschaftspopularisierung im 19. Jahrhundert: Bürgerliche Kultur, naturwissenschaftliche Bildung und die deutsche Öffentlichkeit, 1848–1914.* Munich: Oldenbourg.

Dierig, S., Lachmund, J., and Mendelson, J. A. (eds.) (2003). *Science and the City* (*Osiris* 18). Chicago: University of Chicago Press.

Dietrich, L., and Rieke-Müller, A. (1998). *Carl Hagenbeck: Tierhandel und Schaustellungen im deutschen Kaiserreich.* Frankfurt: Peter Lang.

Dreesbach, A. (2005). *Gezähmte Wilde: Die Zurschaustellung "exotischer" Menschen in Deutschland 1870–1940.* Frankfurt am Main: Campus.

Driver, F., and Gilbert, D. (eds.) (1999). *Imperial Cities: Landscape, Display and Identity.* Manchester: Manchester University Press.

Farías, I., and Stemmler, S. (2012). Deconstructing 'Metropolis': Critical Reflections on a European Concept. In: Brantz, D., Disko, S., and Wagner-Kyora, G. (eds.), *Thick Space: Approaches to Metropolitanism.* Bielefeld: Transcript, 49–66.

Foucault, M. (1970). *The Order of Things: An Archaeology of the Human Sciences.* New York: Vintage.

Frisby, D. (2001). *Cityscapes of Modernity: Critical Explorations.* New York: Polity Press.

Gandy, M. (2003). *Concrete and Clay: Reworking Nature in New York City.* Cambridge: MIT Press.

Geertz, C. (1973). *The Interpretation of Cultures.* New York: Basic Books.

Ginthör-Weinwurm, I.-A. (2008). Die plastische Fassadengestaltung des Naturhistorischen Museums in Wien: Eine Palastwand der Evolution. Diploma thesis, University of Vienna.

Hanson, E. (2002). *Animal Attractions: Nature on Display in American Zoos.* Princeton: Princeton University Press.

Harding, L. (2000). *Elephant Story: Jumbo and P.T. Barnum Under the Big Top.* Jefferson: McFarland.

Heynen, N., Kaika, M., and Swyngedouw, E. (eds.) (2006). *In the Nature of Cities: Urban Political Ecology and the Politics of Urban Metabolism.* London: Routledge.

Hunt, L. (2014). *Writing History in the Global Era.* New York: W.W. Norton.

Jovanović-Kruspel, S., and Schumacher, A. (2017). *The Natural History Museum: Construction, Conception & Architecture.* Vienna: Naturhistorisches Museum.

Klös, H.-G., Klös, U., Strehlow, H., and Synakiewicz, W. (1990). *Der Berliner Zoo im Spiegel seiner Bauten 1841–1989: Eine baugeschichtliche und denkmalpflegerische Dokumentation über den Zoologischen Garten Berlin.* Berlin: Heenemann.

Köstering, S. (2003). *Natur zum Anschauen: Das Naturkundemuseum des Deutschen Kaiserreichs.* Cologne: Böhlau.

Kretschmann, C. (2006). *Räume öffnen sich: Naturhistorische Museen im Deutschland des 19. Jahrhunderts.* Berlin: Akademie Verlag.

Large, D. C. (2000). *Berlin.* New York: Basic Books.

Latour, B. (1991). *We Have Never Been Modern.* Cambridge: Harvard University Press.

Lefebvre, H. (1991). *The Production of Space.* Oxford: Blackwell.

Levin, M., Forgan, S., Hessler, M., Kargon, R., and Low M. (2010). *Urban Modernity: Cultural Innovation in the Second Industrial Revolution*. Cambridge, MA: MIT Press.

Macdonald, S. (ed.) (1998). *The Politics of Display: Museums, Science, Culture*. London: Routledge.

Matejovski, D. (ed.) (2000). *Metropolen. Laboratorien der Moderne*. Frankfurt am Main: Campus.

Metzler, S. (2007). *Theatres of Nature: Dioramas at the Field Museum*. Chicago: The Field Museum.

Miller, D. (1996). *City of the Century: The Epic of Chicago and the Making of America*. New York: Simon and Schuster.

Miller, I. J. (2013). *The Nature of Beasts: Empire and Exhibition at the Tokyo Imperial Zoo*. Berkeley: University of California Press.

Mothes, W. (2013). *Von Bären und Hasen: Dioramen in deutschen Naturkundemuseen*. Mannheim: Edition Panorama.

Nyhart, L. (2009). *Modern Nature: The Rise of the Biological Perspective in Germany*. Chicago: University of Chicago Press.

Osterhammel, J. (2015). *Transformation of the World: A Global History of the Nineteenth Century*. Princeton: Princeton University Press.

Pacyga, D. (2009). *Chicago: A Biography*. Chicago: University of Chicago Press.

Palló, G. (2012). Scientific Nationalism: A Historical Approach to Nature in Late Nineteenth-Century Hungary. In: Ash, M., and Surman, J. (eds.), *The Nationalization of Scientific Knowledge in the Habsburg Empire 1848–1918*. Basingstoke: Palgrave Macmillan, 102–112.

Poliquin, R. (2012). *Breathless Zoo: Taxidermy and the Culture of Longing*. Park City: Penn State University.

Quinn, S. C. (2006). *Windows on Nature: The Great Habitat Dioramas of the American Museum of Natural History*. New York: Harry N. Abrams.

Rader, K., and Cain, V. (2014). *Life on Display: Revolutionizing U.S. Museums of Science and Natural History in the Twentieth Century*. Chicago: University of Chicago Press.

Reif, H. (2012). Metropolises: Histories, Concepts, Methodologies. In: Brantz, D., Disko, S., and Wagner-Kyora, G. (eds.), *Thick Space: Approaches to Metropolitanism*. Bielefeld: Transcript, 31–48.

Riedl-Dorn, C. (1998). *Das Haus der Wunder. Zur Geschichte des Naturhistorischen Museums in Wien*. Vienna: Holzhausen.

Rosenthal, M. (2003). *The Ark in the Park: The Story of Lincoln Park Zoo*. Urbana: University of Illinois Press.

Rothfels, N. (2002). *Savages and Beasts: The Birth of the Modern Zoo*. Baltimore: Johns Hopkins University Press.

Scheersoi, A., and Tunnicliffe, S. D. (eds.) (2019). *Natural History Dioramas: Traditional Exhibits for Current Educational Themes*. Cham, Switzerland: Springer.

Schneer, J. (1999). *London 1900: The Imperial Metropolis*. New Haven: Yale University Press.

Schrock, A. (2019). Why Do We Talk about Cities as Laboratories? https://medium.com/@aschrock/why-do-we-talk-about-cities-as-laboratories-c3d70ff7244f. Download 15 August 2019.

Schwarz, W. M. (2001). *Anthropologische Spetakel: Zur Darstellung "exotischer" Menschen in Wien 1870–1910*. Vienna: turia + kant.

Spary, E. C. (2000). *Utopia's Garden: French Natural History from Old Regime to Revolution*. Chicago: University of Chicago Press.

Supuković, N. (2015). Ecce Animal: Zur Wissenschaftspopularisierung im k. k. Naturhistorischen Hofmuseum und Naturhistorischen Museum Wien am Beispiel von Museumsführern, 1889–1928. M.A. thesis, University of Vienna.

Thackray, J. C. (2013). *Nature's Treasurehouse: A History of the Natural History Museum*. London: Natural History Museum.

Thorsen, L. E., Rader, K. A., and Dodd, A. (eds.) (2013). *Animals on Display: The Creaturely in Museums, Zoos, and Natural History*. University Park: Pennsylvania State University Press.

Wessely, C. (2008). *Künstliche Tiere. Zoologische Gärten und Urbane Moderne*. Berlin: Kadmos.

Yanni, C. (1999). *Nature's Museums: Victorian Science and the Architecture of Display*. Baltimore: Johns Hopkins University Press.

3 Periphery and Metropolis

Some Historiographical Reflections on the Urban History of Science

Oliver Hochadel

Introduction

It is no secret that the biography of a historian might be useful for understanding his/her approach, his/her perspective, and his/her ideas. I had lived in a metropolis (Vienna) for 11 years, but then moved to the periphery (Barcelona) in 2007. A few years later I turned to the urban history of science, very much marked by my geographical location. In this chapter, I take the liberty of using my own research to review the recent historiography on the urban history of science and interurban knowledge exchange.

First of all, I would like to stress that 'my research' in this field is in fact a fundamentally collaborative work. The four volumes on science and the city (Hochadel and Nieto-Galan 2016a; Girón et al. 2017; Hochadel and Nieto-Galan 2018; Gantner et al. 2020) were all co-edited, so there are five other editors involved. A total of 50 scholars authored the 42 chapters in these volumes, some of them contributing chapters to more than one volume, not to speak of the numerous colleagues (commentators, reviewers, and so on) who contributed in important ways to these four edited volumes.

Talking so profusely about my own published work allows me to reflect upon the learning processes co-editors and authors went through. How our understanding of the subject matter evolved is often not fully evident in the finished product, given all the exchanges and debates that went on between the scholars involved in the years of elaborating the edited volumes. The idea of this chapter is not so much to detail the content of our publications, but rather to discuss the conceptual challenges in devising a theoretical framework for them. In more concrete terms, and in order to focus the discussion more specifically on the topic of the present volume, this chapter will reflect upon the terms 'periphery' and 'metropolis' and their changing and often contested meanings. In order to save some ink, I will not put these terms in inverted commas. I shall argue that these terms are best understood as actors' categories and as relational. Their meanings are very much defined by their respective opposites and are thus highly context-dependent.

The four edited volumes mentioned previously will serve as structuring elements of this chapter. I shall try to describe a movement from (1) an individual city such as Barcelona to (2) a comparative approach between cities, and move from there to an (3) entangled history (*histoire croisée*), and eventually (4) to a transurban perspective. The chapter ends with a discussion of a new approach named global urban history (5).

1.   The Spatial Turn in the History of Science— The Barcelona Book

The so-called spatial turn in the history of STM (science, technology, and medicine) is by now well established (Livingstone 2003). It even gave rise to a new field, the historical geography of science (Livingstone and Withers 2011). To focus on the concrete physical space in which knowledge is generated has become a fruitful approach in numerous studies in the past 20 years or so. By now it is widely accepted that scientific knowledge bears the imprint of its site of production; the intriguing question for the historian of STM is what happens to this 'local knowledge' if and when it is transferred to other contexts and gets adapted in the process (Golinski 1998).

This spatial turn has not only placed concrete spaces or institutions such as the laboratory (Shapin 1988) or the museum (Forgan 2005) at the center of attention, but also larger and far less defined spaces such as the city. The urban space represents in fact both scale and site, as Withers and Livingstone (2011, 5–6) argue: "Thinking about cities as sites of and for science might seem to sanction imprecision. . . . This is not so." Rather, the historian must now examine how "scientific knowledge has to travel between and within city sites and out from them to different audiences."

Historians of STM have postulated a dialectical relationship between science and urban spaces. On the one hand, science, technology, and medicine shape the urban space, its institutions and infrastructures, including vital urban features such as sewage systems, public transport, and public health. On the other hand, the urban space enables, facilitates, or obstructs the unfolding of STM, the production, circulation, and implementation of knowledge. As the editors of a special issue of *Osiris* on "Science in the City" put it: Rather "than a passive container of institutions and practices, urban space was a complex material and symbolic environment that was shaped by—and that in turn shaped—institutions in historically specific ways" (Dierig et al. 2003, 5). A very recent and, in fact, splendid example for pursuing this research agenda is Tanya O'Sullivan's book on late Victorian Dublin (2019).

Surveying the historiography on science and the city, we, historians of STM from Barcelona, clearly felt that the attention of our colleagues had been directed nearly exclusively toward metropoles and industrial towns. Much research had focused on large cities such as London, Paris, Berlin, and Chicago (Levin et al. 2010). This book focuses mainly on Vienna.

Already some decades ago historians of STM looked at the scientific culture of regional or provincial cities (for example Kargon 1978; Inkster and Morrell 1983; Morrell 1985). Yet their focus was largely restricted to Great Britain (but see Nye 1986, on French provincial cities), where the relevance of cities such as Manchester and even Bradford is obvious due to their seminal role in the industrial revolution. Only very recently Coen (2018, in particular 9, 73, 97, 99, 290) discussed Vienna-based research networks in climatology and other fields, emphasizing the dialectic relationship between metropolitan and provincial science.

For our specific interest it is important to note that the duality of metropolis and periphery (or rather province/provincial) was coined historiographically in this particular British context. Inkster and Morrell (1983) entitled their edited volume *Metropolis and Province. Science in British Culture, 1780–1850*. And it is certainly no coincidence that a current research project lead by Rebekah Higgit is called "Metropolitan Science. Places, Objects and Cultures of Practice and Knowledge in London, 1600–1800" (https://metsci.wordpress.com).

The towering presence of London structures the terminology historians of STM deploy: metropolitan science versus science in the provinces and also first city versus second cities. Clearly, the discussion in the history of STM reflects analogous discussions in other fields, for example in urban history. The book *London 1900: The Imperial Metropolis* (Schneer 1999) is an instructive example. Talking about the metropolis London means by definition to take the colonial dimension of the British Empire into account. Paris was, of course, the other world city at the time, a reference point for other cities but also for political and social movements in a number of respects, as we shall see in the following.

In this large shadow of London and Great Britain and only in recent years, intriguing studies on scientific culture in nineteenth-century German cities have appeared. Dresden (Phillips 2012) and Frankfurt am Main (Sakurai 2013) were marked by their bourgeois elites and their cultural and scientific institutions, not by industrialization. One may also mention Deborah Coen's magisterial *Vienna in the Age of Uncertainty* (2007), although the protagonist of the book is not the city itself but the Exner family, a dynasty of scientists who lived and worked there. Curiously, these three books about German-speaking cities were written by scholars trained in the Anglo-American academic sphere.

Given this historiographical panorama, it may be understandable that historians located in the periphery felt marginalized, and they have reacted. The truth is that I only started to understand this discontent after moving from Vienna to Barcelona. In 1999, historians of STM from Greece, Spain, Portugal, Hungary, Sweden, Ireland, Turkey, and other countries often considered peripheral launched the research network STEP (Science and Technology on the European Periphery). Scholars of STEP asked what historians of science and technology located on the periphery should actually study. Is there anything that can be learned by doing history of STM in countries with apparently few famous scientists and prestigious institutions? After 20 years of research, the answer is clearly yes. Knowledge is not created exclusively in the center and then passed on to the periphery. Historians of science not only from STEP have repeatedly emphasized that knowledge is changed in the act of communication. In new contexts historical actors appropriate and adapt ideas, practices, and material objects such as instruments in their own, often innovative ways (for a synthesis of the STEP agenda and its results see Diogo et al. 2016).

The initial idea that eventually led to the edited volume on the urban history of science in Barcelona (Hochadel and Nieto-Galan 2016a) was to pursue the STEP agenda with respect to science in the city, and hence to ask: Is there something about the urban peripheries that allows us to learn something we cannot learn by studying the scientific culture of London or Paris? We analyzed the production, communication, and appropriation of knowledge, its connection to the fabric of the city and the everyday life of its inhabitants around 1900. The ten case studies try to pursue the spatial approach, to focus on concrete localities in the city: In which quarters, in which institutions and networks can specific practices in STM be identified? And what kind of experiences did the city dwellers make when interacting with scientists, engineers, and physicians around 1900? Who were the elites and the experts?

In asking these questions, we always tried to include the audiences of STM, in the broad sense of the term, that is to say not only the visitors of museums but also, for example, medical patients and users of new technologies (radios, electrical appliances). The spatial approach yielded interesting results: Many of the new amusement parks (based on new "technologies of entertainment") were built on the margins or just outside of Barcelona (Sastre and Valentines 2016). A large number of new private clinics clustered in the recently built (and very bourgeois) Eixample district (Zarzoso and Martínez-Vidal 2016).

It thus became clear that practices in STM in Barcelona around 1900 had 'class' written all over them. These practices were by no means ideologically neutral and played an important role in the agendas of political and social movements. For the bourgeois, conservative, Catholic, and also Catalanist segments of Barcelona's society these practices served to maintain their hegemony, not only in a political but also in a cultural sense. STM were vital to expose their world view—rejecting the theory of evolution, for example—and the moral values attached to it. At the same time, diverse currents represented by republicans, freethinkers, anarchists, and even spiritists and theosophists tried to use STM to further their counter-hegemonic political and social agenda.

These results led to new questions. While finalizing the book we wondered: Had Barcelona in a sense become herself a scientific metropolis?

At the time we did not pursue this issue, and it remains an open research question. A few tentative remarks must suffice here. If the criterion for metropolis status is an increase in institutions in STM, then one may answer this question affirmatively. A few founding dates: the first public museum (founded 1882, later morphing into a natural history museum), a municipal laboratory dealing with hygiene (1887), a zoological garden (1892), the *Institució Catalana d'Història Natural* (1899), the *Institut d'Estudis Catalans* (1907). Many more institutions, collections, and associations could be named here. This institutional 'densification' is, however, quite typical for European cities around 1900. New networks of practitioners in STM emerged, within the urban space but also connecting institutions on a national and international level. The time frame of the book was specifically chosen to cover the four decades between two international exhibitions that took place in 1888 and 1929 that were meant to put Barcelona on the global map.

Asking whether Barcelona became a metropolis around 1900 also points to the dialectical relationship of the urban space with its environment. In particular, numerous exchanges of the city with the Catalan countryside emerged in many of the contributions to the volume. Naturalists explored and described the Catalan fauna, flora, and geology. A variety of allochthonous species of fish were bred in the Barcelona zoo and then introduced into Catalan rivers and lakes in order to repopulate them. Yet we did not pursue the question of this mutual transformation of city and countryside in a more systematic manner (but see Valls 2019). Clearly more research combining urban and environmental history of STM is called for. The magisterial model is *Nature's Metropolis. Chicago and the Great West* (Cronon 1991).

Yet to resume our experience with the Barcelona book: As we applied the spatial approach and thus could highlight the ideological resonances of practices in STM, we realized that there was no need for the concept of periphery to produce a rich account of Barcelona's urban culture of science.

2. The Periphery Problem—Urban Histories of Science

How exceptional or typical was the urban history of science of Barcelona? This was the guiding question in our subsequent project that eventually culminated in the publication of another edited volume (Hochadel and Nieto-Galan 2018). *Urban Histories of Science. Making Knowledge in the City, 1820–1940* examined eight European cities: Athens, Barcelona, Budapest, Dublin, Glasgow, Helsinki, Lisbon, and Naples as well as Buenos Aires.

All the titles of the workshops and conferences we had organized to prepare the book featured the term periphery, for example: "Urban Peripheries? Science in 'second cities' around 1900". Yet in these meetings this concept reliably provoked a discussion about its utility, despite the fact that we put it in quotation marks. Some colleagues suggested that the use of the term might perpetuate stereotypes about the alleged 'backwardness' of the cities in question. We were wondering whether there were conceptual alternatives to the term periphery. A term used by historians (including ourselves) has been "second cities" (Umbach 2005 about Barcelona and Hamburg; Adelman 2018 about Dublin, also see Ewen 2008). Yet this expression entails the same problem, presupposing a hierarchy (of size or of relevance) between 'first' and 'second' cities. On occasion we even spoke of the 'non-metropolis' in a rather desperate attempt to define the kind of city we were dealing with.

In the end we felt that the term periphery was not helpful but rather a burden. Just as we had done earlier with the Barcelona book, we realized that this question periphery-center might deflect our attention from the more interesting issues at stake. So striking the term from the title felt rather like a liberation. *Urban Histories of Science* sounded much better to us. To be sure, the term periphery did matter, but only as an actors' category. It was used at the time in the late nineteenth and early twentieth century, along with a host of similar terms from the same semantic field, to lament the alleged or perceived backwardness of a city. In fact, it became a rhetorical resource to plead for redoubling efforts to 'catch up' with more advanced cities or with 'civilization'.

In this sense, that is as a rhetorical resource, the term metropolis surfaced in some of the chapters of *Urban Histories*, twice even in the title. Athens was little more than a village at the beginning of the nineteenth century, yet its Bavarian rulers and Greek intellectuals aspired to make it a metropolis again (Rentetzi and Flevaris 2018), even alluding to the original meaning of metropolis from antiquity as a mother city to the colonies. In the mid-nineteenth century, Glasgow and its booming ship industry tried to claim a place in the hierarchy of British cities, the London metropolis, and the industrial cities farther north. In this context, the leading engineer W. J. M. Rankine launched the concept of Glasgow as "Metropolis of Mechanics" (Marsden 2018).

Periphery—metropolis, backwardness—modernity or provincial are European binaries that were semantically close and were often used interchangeably in the discourses of urban elites around 1900. Urban historians have amply analyzed these discourses. For example, Eastern European cities (the new capitals after the collapse of the empires) have been described as engaged in *Races to Modernity. Metropolitan Aspirations in Eastern Europe, 1890–1940* (Behrends and Kohlrausch 2014).

What the sum of case studies in *Urban Histories* brought to the fore is that STM played a crucial role in the vision of these nine cities of how to modernize themselves. State-of-the-art astronomical observatories or marine laboratories, zoological gardens with international vibrancy, highly publicized meetings of scientific societies or a well-equipped public health system fighting the spread of tuberculosis were tangible markers of this coveted modernity.

In *Urban Histories* we tried to show the quite different notions of modernity that these cities developed for themselves. Modernity became locally varied and was never homogeneous. More than that: as already mentioned with respect to Barcelona, different social groups (such as bourgeois elites or anarchists) constructed diverging images of their own city. A helpful concept in order to capture this diversity is the idea of "multiple modernities" first formulated by sociologist Shmuel Eisenstadt (2000; for the limits of this approach see Conrad 2017, 60–61). Eisenstadt criticized the idea in Western thought that there was only one "cultural program of modernity" (2000, 1) that was supposed to have spread from Europe to the rest of the world. For him modernity was, rather, a set of "multiple institutional and ideological patterns . . . carried forward by specific social actors . . . pursuing different programs of modernity, holding very different views on what makes societies modern" (2000, 2). If we apply this idea of multiple modernities to urban history, we might liberate our narratives from the straightjacket that equates urban modernity with the metropolis, without forgetting that the actors saw things differently.

3.   From Comparison to *histoire croisée*: Barcelona, Buenos Aires, and Beyond

Comparison has always been a useful tool for the historian in order to contextualize a specific local case study. The differences that emerge in the comparison help to identify unique features of the case at hand. As described previously, in *Urban Histories* we tried to highlight the differences between a number of cities as regards the role of STM in their modernity discourses.

Yet the comparative approach clearly has its limits (for a general critique see Conrad 2017, 40). As has been pointed out by a number of historians, comparison carries the danger of essentializing the items in question. As Michael Werner and Bénédicte Zimmermann (2006, 37) stated, in such comparisons "the reference points of the analysis are not questioned as such." These influential proponents of *histoire croisée*, called entangled history in English, argue that local, national, and transnational levels cannot be separated. The objects of research can only be grasped fully if they are understood as mutually constituting one another. Historian of science Lissa Roberts (2009, 22), for example, wrote: "it becomes more difficult to rest content with a comparative approach that begins by identifying similarities and differences rather than confluences and mutually formative interaction."

While such calls for a historiography that focuses on interaction have become frequent in recent years, instructive examples are arguably less numerous. The reason is simple: It is very challenging actually to construct and tell interurban, transnational, interconnected histories, or a *histoire croisée*, whatever we prefer to call them. The historian faces the task of a juggler keeping several balls in the air at once. How exactly do the objects of research constitute each other?

I am now part of a Spanish-Argentinian research group that looks at the multidirectional exchanges of objects, people, and ideas between Barcelona and Buenos Aires. A first product of this group was the edited volume *Saberes transatlánticos. Barcelona y Buenos Aires: conexiones, confluencias, comparaciones (1850–1940)* (Girón et al. 2017) (*Transatlantic knowledges. Barcelona and Buenos Aires: connections, confluences, comparisons (1850–1940)*). As the title indicates, our nine case studies look at exchanges between Barcelona and Buenos Aires in a variety of ways, including more traditional methods such as comparison and contrast.

Yet some of the case studies of this book could only be told as *histoires croisées*, examples of which I shall now briefly sketch. One example is the radical political vision of the Argentine intellectual (and later president) Domingo Sarmiento (1811–1888). Since the mid-1840s Sarmiento presented Madrid and Barcelona in a strictly antithetical way. Spain stood for the old aristocratic rule, superstition and backwardness, in short "barbarism", while Catalonia represented the modern liberal values of education, entrepreneurship, and progress, in short "civilization", to use Sarmiento's key terms. This polemical juxtaposition needs to be understood within Sarmiento's political vision for Argentina, in which Buenos Aires embodied the values he championed and Barcelona served as a crucial point of reference. The Catalan Luis Ricardo Fors de Casamayor (1843–1915) was initially a political ally of Sarmiento, but then he embraced much more radical democratic-republican thinking. Fors criss-crossed the Atlantic for many years, moving between Barcelona and Buenos Aires but also setting up camp in Montevideo, Asunción, Madrid, Lisbon, and Seville before finally settling in La Plata, a new model city about 60 kilometers away from Buenos Aires. In his lifelong struggle for his pedagogical and political ideals, he had to take refuge from persecution more than once. For the Catalan Fors the South American city was a true laboratory for progressive ideas (Vallejo 2017).

Another example for such a transatlantic biography is Víctor Grau-Bassas (1847–1918). The Spanish naturalist, physician, and museum curator was born in Las Palmas (Canary Islands), studied medicine in Barcelona (where his family had come from), and, like Fors, eventually settled in La Plata. Grau-Bassas had worked in the Museo Canario in Las Palmas, but due to personal circumstances emigrated to Argentina, finding work at the museum in La Plata. There he acted as an intermediary between the two museums, and thus the research networks on both sides of the Atlantic, facilitating the circulation of anthropological objects and the exchange of scientific knowledge (Betancor Gómez 2017).

The close ties of the anarchist movement between Barcelona and Buenos Aires around 1900 provide quite a different example of a *histoire croisée*. It was very common for anarchists to go back and forth between the two cities, promoting anarchist journals or simply seeking refuge from police persecution. These exchanges were generally clandestine and thus were marked by repression. Hence the networks of the anarchists were fragile and transient. Nevertheless, they were able to produce vivid discussions, for example about the significance of terms such as 'race' and 'national roots'. To be sure, anarchism is by definition an internationalist movement. Yet particularly in the urban space of Buenos Aires, characterized by its large immigrant population and a striking ethnic and linguistic admixture, these terms were widely discussed by anarchists (Girón 2017).

Three comments are needed in order to put these *histoires croisées* into a larger context:

1. What we labeled "saberes transatlánticos" (transatlantic knowledges) refers not only to the history of STM but also to the history of literature and political thought, intellectual history, and the history of education. This reflects in a sense the Argentine scholarly community, in which there is no neat separation between these fields (see for example Terán 2000).

2. One recurring critical question we encountered when presenting our book and the corresponding research program was this: Why would you choose Barcelona and Buenos Aires? Is this not rather arbitrary? We then pointed out that the circulation of people, ideas, practices, and objects between the two (mainly) Spanish-speaking port cities was indeed very intense in the later part of the nineteenth century and well into the twentieth century. Examples abound, also in the fields of literature, spiritism, psychology, and so on. Yet, and maybe more to the point: In the three articles just mentioned the *histoire croisée* unfolds between more than just two geographical poles, that is Barcelona and Buenos Aires, but among three or more cities. In this sense the initial focus on two urban spaces is a heuristic approach that helps to identify connections, but is always only a first step. The fact that other urban spaces become part of the *histoires croisées* is, of course, to be expected and sheds light on the interurban networks that connect them.

3. To talk about interurban networks and exchanges should not obscure the fact that there remains an urban hierarchy of sorts. In the case of Grau-Bassas and the two museums in Las Palmas and La Plata, Paris and the leading French anthropologist Paul Broca (1824–1880), who worked there, remain a crucial point of reference and constitute the ultimate authority. Similarly, although in a completely different field, reconstructing the networks of the anarchist movement around 1900 shows that Paris and London were the crucial global hubs of

anarchist thought. Many 'high impact' publications, for example by Élisée Reclus and Piotr Kropotkin, were published in those two metropoles and started to circulate—mostly in texts written in French that were then translated—in the transnational networks of the anarchist movement (Girón 2019).

These three quite different *histoires croisées* only provide some spotlights, but they are strong indicators that the urban history of science around 1900 can only be written as interurban history. Such an approach may enable the historian to investigate how ideas, but also actors and objects change when they move to other urban spaces. Therefore, it seems crucial to conceive of this relationship between two cities as a dynamic and dialectical one, without losing sight of the larger urban networks in which they participated.

4.   Interurban Connections: Southern and Eastern Europe

This chapter argues that an important step further in this line of research is to conceptualize cities as network nodes, irrespective of their size, geographical location, or alleged importance in larger interurban networks. Historians of STM interested in the communication and exchanges that connected cities might look for guidance in urban history. I will briefly describe two historiographical tools that I found particularly useful: 'transnational municipalism' and 'the interurban matrix'.

Urban historians have forged the concept 'transnational municipalism' in order to understand the circulation of specifically urban knowledge between cities (Hietala 1987; Saunier and Ewen 2008; Kenny and Madgin 2015). Starting in the last decades of the nineteenth century, many city councils sent commissions to other cities all over Europe, or even across the Atlantic, to gather information on how to modernize their own cities. Disregarding national borders, cities had become increasingly aware that they were facing similar problems. In these fast-growing cities issues of hygiene, a functioning sewage system, and sufficient drinking water supply were most pressing. To provide city-dwellers and growing industry with gas and later electricity, and to devise a functioning public transport system (trains, trams, and trolley-buses) were other typical challenges for urban magistrates. They were thus eager to learn from one another and to adapt 'best practices', as we would say today, with respect to how to build and run a modern city. These study trips were mostly dedicated to questions of public health and urban planning, but also to how to found or reform specific institutions, such as museums or zoological gardens. World's fairs and other international exhibitions also served as show cases for how a modern city should look.

As the research on transnational municipalism has shown, city councils were well aware that the metropolis might not always have the proper solution for their own specific urban problems. As urban planners quickly understood, cities of comparable size might have developed concepts that would fit their own specific needs much better. Thus, city governments tried to avoid errors that had been committed elsewhere, or models that had proven problematic (Gantner and Hein-Kircher 2017, 5).

Another useful concept from urban history is the idea of an 'interurban matrix', coined by Nathan Wood while pursuing his research on Cracow around 1900. From the mid-nineteenth century onwards, both the number of newspapers and literacy rates increased vastly all over Europe. Wood argues that the ubiquitous daily press made its readers aware that they shared very similar urban experiences with inhabitants of many other European cities. Urban readers of newspapers were bound together by a set of experiences, aspirations, and values which found its most condensed form in the expression "European civilization", meaning modern transport systems, public health services, and cultural institutions such as the opera or the theatre (Wood 2006). Wood's book on Cracow (Wood 2010) is entitled *Becoming Metropolitan*. By that he means "the process of adaptation to modern urban life, in such a way that one was conscious, even if only indirectly, of one's participation in what contemporaries frequently termed 'modern urban civilization'" (Wood 2010, 12–13). Recently Guarneri (2017) made a similar argument for US cities in *Newsprint Metropolis: City Papers and the Making of Modern Americans*.

A few years ago I embarked on a cooperation with Eszter Gantner and Heidi Hein-Kircher, both historians of Central Eastern Europe. They had coined the concept of "emerging cities" (Gantner and Hein-Kircher 2017), arguing that Eastern European cities should be understood as agents in their own right. In the late nineteenth and early twentieth centuries cities such as Prague, Budapest, Lviv (formerly Lemberg), Cluj-Napoca (formerly Klausenburg), and Chernivtsi (formerly Czernowitz) developed into modern regional metropolises, and in some cases into national capitals after 1918. Belfast, too, had been labeled an emerging city (Purdue 2013). These cities had to straddle between the modernization of the urban space, on the one hand, by following examples from abroad, and forging a local and often also national identity on the other hand—two interdependent but also often contradictory agendas.

The idea of our common project was to focus on two quite distinct peripheries, Eastern and Southern Europe. Did they have a reference metropolis? Would we find similar patterns of knowledge exchange? Would we even find instances of exchanges among the peripheries?

In order to answer these questions, we used the previously mentioned tools from urban history, transnational municipalism, and the interurban matrix, but also the concept of "knowledge in transit" (Secord 2004) from the history of STM. Our 12 case studies in this volume (Gantner et al. 2020) attempt to show that the knowledge that circulated in this transnational space was permanently altered, combined, hybridized, and adapted (although not always successfully) to fit the specific needs of an emerging city. In practice the solutions applied were often idiosyncratic and highly eclectic mixtures of different recipes. As Shane Ewen (2015, 124) already observed: "Foreign models and innovations never simply materialized as static things." The chapters in our volume pay particular attention to the large spectrum of media that gather, convey, and discuss this genuinely urban knowledge: traveling experts, growing transnational networks, publications of all sorts (from newspapers to handbooks), exhibitions, fairs, and prize competitions.

Here are some examples from this volume (Gantner et al. 2020) for the circulation of urban knowledge: Czech experts pondering how to reconstruct Prague's Jewish ghetto in studying urban-renewal practices in cities such as Toulon, Naples, and Florence; a close collaboration between art historians and politicians from Barcelona and Bucharest not only on Romanic architecture but also on nation building; sanitary reform in Imperial Moscow copying crucial ideas from the sewage system of Memphis, Tennessee; the oscillation of plans for a Barcelonese tuberculosis dispensary between Soviet and Italian Fascist best practices (before planners opted for a 'Mediterranean' model); a Hungarian architect and interior designer leaving his mark on Detroit and Mexico City; a French envoy drawing inspiration from animal parks in Northern and Eastern Europe in order to reform the zoo in Paris; the early Barcelonese automobile industry and its considerable impact in France and the United States; an Italian industrialist turning to models on the eastern side of the Iron Curtain to build a science museum in Milan in the early 1950s.

To stress this point once more: In any of these cases the local urban context, its specific political, social, economic, and cultural characteristics, is of crucial importance for this transfer of urban knowledge. Yet clearly, to ask about the circulation and appropriation of this kind of applied urban knowledge allows us to paint a much richer picture of the complex dynamics of the urbanization processes around 1900. What emerges is an interurban and by definition transnational space, tied together by contacts among experts (architects, urban reformers, engineers, business men, physicians, and other specialists), but also by a vast range of publications including the printed press.

Again I make three points to conclude this section:

1. The concept of emerging cities evokes far more positive connotations than periphery or second cities, because it suggests agency. But once more, just like with periphery in *Urban Histories* (Hochadel and Nieto-Galan 2018), discussion with colleagues and anonymous reviewers (of the book proposal) led us to strike the qualifier from

the title. Some of our cities are rather "old cities" (for example Moscow, Prague, and Barcelona). Yet more importantly, even with a positive adjective such as emerging there is always the risk of reproducing narratives of center-periphery that the volume pretends to challenge. The original title for the edited volume was *Interurban Knowledge Exchange. Emerging Cities in Southern and Eastern Europe, 1870–1950*, which we eventually condensed into the more neutral *Interurban knowledge exchange in Southern and Eastern Europe, 1870–1950*.

2. What does this mean for the center-periphery model and the idea of the highly influential metropolis? Again, just as we observed in the Barcelona-Buenos Aires case: to highlight the numerous interactions between cities generally not considered metropoles and sketching a vast group of urban spaces engaged in exchanging knowledge and best practices does not imply a symmetrical panorama. It simply provides a far more nuanced picture, in which all kinds of interurban exchanges occurred. This includes the metropolis as a powerful urban model. Alphand's urbanistics manual *Les Promenades de Paris* (1867–1873) provides an instructive example for a "centrifugal transfer" of knowledge from Paris to Eastern Europe (Stühlinger 2017, 47). The strong influence of Haussmann's Paris and of the Vienna Ringstrasse (and of the ideas of Austrian urban planners such as Camillo Sitte and Otto Wagner) on the new Central and Eastern European capitals is well documented (Gunzburger Makas and Damljanovic Conley 2010a). Yet they also stress that these models "were adapted within a specific physical and historic context", and add that "[m]ost cities were actually transformed in more piecemeal fashion" (Gunzburger Makas and Damljanovic Conley 2010b, 18–19). The reference metropolis for the Eastern European cities varied very much between Vienna, Paris, and Berlin, depending on the area of reform and the given political constellations. The city council of Lemberg (Lviv), for example, searched for best practices in sanitary reform in Germany rather than Vienna, consciously distancing themselves from their own capital.

3. Were there exchanges of best practices between Southern and Eastern Europe? We came upon a number of rather unexpected connections (see previous page) that document the circulation of knowledge and practices among peripheries, bypassing in a sense the metropolis. Yet similarly to the experience with the focus on the connections between Barcelona and Buenos Aires, this focus turned out to be too narrow. Interurban exchanges took place in all directions across Europe. And at least in some of the chapters in these volumes, cities in North and South America also made an appearance. Where do we go from here?

5. Global Urban History

Reconstructing the dynamic exchanges between cities on a European or even on a global scale is clearly a fruitful approach, as some recent studies demonstrate. John Griffith (2009, 581), for example, looking at municipal networks in the British Empire before 1939, states that there was "at no stage a limited 'one-directional' flow of knowledge from Britain to the outer reaches of the Empire". It becomes essential to look well beyond European and North American cities, as Nora Lafi (2008, 35) observes: "If discussions largely remain centred within the limits of the Western World as it is generally perceived, and the circulation of urban knowledge read through the lens of a limited geographical horizon, there can be no true global history."

Some urban historians have suggested that postcolonial theory may be instrumental in deconstructing the claim of Western civilization as the one and only root of modernity. This theory intends to show ubiquitous 'othering' in a binary discourse of 'modern/European' versus 'underdeveloped', 'civilized' versus 'uncivilized', 'advanced' versus 'backward' or periphery versus metropolis that marks—often more implicitly than explicitly—urban history (Dibazar et al. 2013, 655). While postcolonial theory is certainly useful for criticizing the othering of non-Western cities, the question that poses itself is what could substitute for this problematic narrative of the triumph of 'Western Civilization'. The answer some historians have proposed is global history in a new, decentered sense, as opposed to traditional universal history. What seems opposed at first, the global and the urban (local) could and should be fused, as urban historians have argued. A splendid example for a successful combination of these approaches is Michael Goebel's *Anti-Imperial Metropolis* (2015). In this book, Goebel describes how inter-war Paris may be seen as the crucial space for the politicization of numerous intellectuals and future political leaders from Asia, Africa, and Latin America, and thus as the cradle of "third-world nationalism".

The benefits could be mutual, as urban historian Carl Nightingale (Global Urban History 2016a) suggests: "World history can teach urban historians to think more about connections, to encourage us to take the risk of leaping beyond our archive into the synthetic, and to be more geographically inclusive, especially of cities in the Global South. Urban historians can teach world historians how to be messier and richer with their understanding of the complexities of urban nodes." Urban historian Shane Ewen (Global Urban History 2016b) thus pleads: "we can learn much from global history, in knitting together our local studies into something that is more than the sum of its parts, and in saying something more general about the nature of urban life."

The point of combining urban history and global history is that "cities are the locations par excellence where global history takes place"

(Exenberger and Strobl 2013, 12). This collaboration might not only help to bid good-bye to an outdated focus on the metropolis, but also to overcome "methodological nationalism" (Middell and Naumann 2010, 155). In recent years 'transnational' or 'transurban' histories have become buzz-words. "European cities are seen as nodes in communication networks that created a transurban public sphere," and this helps to question whether the emerging nation-state of the nineteenth century should be the main frame of historical analysis (Møller Jørgensen 2017, 558).

It is too early yet to say what will become of this global urban history. It is not a fully fledged field, but rather a research agenda pursued by urban historians such as the already mentioned Carl Nightingale and Michael Goebel. They have formed a "Global Urban History Group" hosting a website including a blog (www.globalurbanhistory.org) and some conferences (2018ff.). This approach is part of a much larger shift toward a new global history that knows how to alternate between different scales.

Whose Periphery? Whose Metropolis?
Some Final Considerations

This chapter has taken the periphery-metropolis opposition as a red thread to discuss the recent historiography on 'science and the city', including a fair dosage of new approaches from urban history. Taking those four edited volumes as a trajectory, we moved from an individual city to a comparative approach, and then on to a perspective that focuses on interurban connections and exchanges. As it turned out, a focus on a specific pair of cities such as Barcelona and Buenos Aires, or two geographic spaces such as Southern and Eastern Europe, may be a useful starting point in order to reconstruct interurban connections. Yet very quickly any geographical limitation will have to give way to complex and multidirectional exchanges between a large number of urban spaces.

In recounting the theoretical framework of each of the edited volumes, this chapter also, but less explicitly moved from a rather disciplinary to an interdisciplinary approach. It started off with a perspective informed by the history of STM looking at a specific space, the peripheral city around 1900. From urban history of science the focus changed to a more applied kind of specific urban knowledge. As we showed, city halls were looking for best practices in order to solve the serious problems of their rapidly growing cities in public health, infrastructure, and urban planning. In the end we cursorily discussed new approaches from global history and how they might be fused with urban history and the history of science. This is obviously not an easy task, but given the historical reality of the interurban space around 1900 and its intrinsic complexity it seems to be the only viable route.

It goes without saying that interdisciplinarity should always include reflexivity. If I may speak from my own experience of co-editing those four volumes: writing this kind of transurban history is quite a challenge. There are several reasons for this. As I noted previously, such research is by definition interdisciplinary and requires from the participating scholars that they seriously engage with neighboring disciplines. Yet it goes further than that. This kind of research is only conceivable as an international project. This means, at least in the cases presented here, engaging seriously with different academic cultures, and not only different disciplines. Such an endeavor is, in our experience, both enriching and challenging. In Argentina, for example, there is no clearly defined community of historians of STM; their academic cosmos is structured differently, their take on "urban topics" shaped by their specific historiographic traditions. Although generalizations are as always treacherous, historians in Eastern European often tend to pursue a very different approach, that we may describe as positivistic. Many of these historians are hesitant about 'Western theories'. Careful negotiations and compromises are needed in order to avoid reproducing hierarchical relationships or serious dissonances. It was a curious experience to see that the dialectical relation between periphery and metropolis lurks in our own professional relationships in the twenty-first century.

Even if the term is placed in inverted commas, as we did in our workshops, the term periphery irritates. As described previously, different sets of co-editors decided not to put periphery or similar terms into the title of our books. Obviously though, the problem cannot be made to disappear by simply eliminating a problematic term from the title. We suggested, instead, to understand periphery (or province) and metropolis primarily as actors' categories. Historical actors used these terms as rhetorical resources. In our cases they mostly served as a call to action in order to modernize, civilize, Europeanize the urban space in order to catch up with the metropolis. The focus on multidirectional exchanges of knowledge should not conceal that the interurban space is not made up of urban knots of equal standing. It is deeply marked by cultural, political, and economic hierarchies. Depending on the case under study—the anarchist sphere, natural history, or urban planning—this may play out very differently. The reference metropolis changes accordingly but, as we have seen, the metropolis as such (be it as a concrete model or an ideal, an intellectual hub, or a space for the socialization of elites from the periphery) does not disappear, despite the decentering historiographical approaches.

Scholars of postcolonial theory, but also urban historians, have pointed out repeatedly that the category of the metropolis (and thus in turn of the periphery) is drenched with significance, with ideologies and politics; these categories bear the marks of the attempt to establish a center of political, economic, and also epistemological power dominating the periphery/province/colony. The term metropolis may take on a host of different meanings depending on specific contexts (Brantz et al. 2012a;

idem., 2012b). In other words: historicizing each case is key. And this must necessarily include our present time as well.

While we struck the term periphery from the titles of our edited volumes, the reader may have noticed that many books cited in this chapter include the term metropolis: *Nature's Metropolis* (Chicago), *The Imperial Metropolis* (London), *Anti-Imperial Metropolis* (Paris), *Becoming Metropolitan* (Cracow), and *Newsprint Metropolis* (US cities). This volume and the conference from which it originated bear the title *Science in the Metropolis* and focus mainly on Vienna, which has served as a reference metropolis in many cases. Sounds good, does it not? In fact, the book market is flooded with titles that use this evocative and iridescent term (for the inflationary use of the term in general see Brantz et al. 2012b, 11).

Not unlike our historical actors or the PR departments for city marketing of today, historians too deploy the term metropolis as a rhetorical resource in order to convince funding bodies or publishers to support and publish their research. Even if it may be impossible to give a succinct definition of metropolis, the term certainly resonates and helps to communicate to non-specialists the supposed relevance of one's own research.

Periphery is certainly a less glamorous term, but I would freely admit that putting it in the title of many of our conferences and occasionally in the title of a publication (Hochadel and Nieto-Galan 2016b) was also a strategic decision, taken in order to attract attention or simply to claim an underexplored niche. As mentioned at the beginning of this chapter, our project began with the discontent of Barcelonese historians of STM about the lack of studies of 'our city' and the dominance of Anglo-American scholarship in the field of science and the city. Steeped in the STEP agenda as we were, we lamented the alleged disinterest of our colleagues in the periphery. To overcome this feeling of being marginalized and to vindicate one's own research may blur the line between the object of study and the historian. There are surely all kinds of ways to reflect upon one's own position while exploring interurban knowledge exchanges. For me, the move from the metropolis to the periphery has been crucial.

Acknowledgments

Research for this chapter was supported by the project "Ciencia y Ciudad. Historia Natural, Biología y Biopolítica en la Urbe Dividida. Barcelona frente a Buenos Aires (1868–1936) (HAR2013–48065-C2-1-P, Ministerio de Economía y Competitividad)," the Proyecto intramural especial "¿Ciencia en la periferia? La cultura científica de ciudades en el Sur y el Este de Europa alrededor de 1900: paralelismos, contextos y redes" (201510I030, CSIC) and the Grup de recerca consolidat i finançat (2017 SGR 1138, AGAUR-Generalitat de Catalunya).

Bibliography

Adelman, J. (2018). Second City of Science? Dublin as a Centre of Calculation in the British Imperial Context, 1886–1912. In: Hochadel, O., and Nieto-Galan, A. (eds.), *Urban Histories of Science*. London: Routledge, 122–140.

Behrends, J. C., and Kohlrausch, M. (2014). Races to Modernity: Metropolitan Aspirations in Eastern Europe, 1890–1940: An Introduction. In: Behrends, J. C., and Kohlrausch, M. (eds.), *Races to Modernity: Metropolitan Aspirations in Eastern Europe, 1890–1940*. Budapest, New York: Central European University Press, 1–20.

Betancor Gómez, M. J. (2017). Una biografía científica atravesando tres ciudades: Víctor Grau-Bassas en Barcelona, Las Palmas y La Plata. In: Girón, Á., Hochadel, O., and Vallejo, G. (eds.), *Saberes transatlánticos. Barcelona y Buenos Aires: conexiones, confluencias, comparaciones (1850–1940)*. Aranjuez: Doce Calles, 133–157.

Brantz, D., Disko, S., and Wagner-Kyora, G. (eds.) (2012a). *Thick Space: Approaches to Metropolitanism*. Bielefeld: Transcript.

Brantz, D., Disko, S., and Wagner-Kyora, G. (2012b). Thick Space: Approaches to Metropolitanism: Introduction. In: Brantz, D., Disko, S., and Wagner-Kyora, G. (eds.), *Thick Space: Approaches to Metropolitanism*. Bielefeld: Transcript, 9–27.

Coen, D. R. (2007). *Vienna in the Age of Uncertainty: Science, Liberalism, and Private Life*. Chicago: The University of Chicago Press.

Coen, D. R. (2018). *Climate in Motion: Science, Empire, and the Problem of Scale*. Chicago: The University of Chicago Press.

Conrad, S. (2017). *What is Global History?* Princeton: Princeton University Press.

Cronon, W. (1991). *Nature's Metropolis: Chicago and the Great West*. New York/London: W.W. Norton.

Dibazar, P., Lindner, D., Meissner, M., and Naeff, J. (2013). Questioning Urban Modernity (Introduction Special Issue). *European Journal of Cultural Studies* 16:6, 643–658.

Dierig, S., Lachmund, J., and Mendelsohn, A. (2003). Introduction: Toward an Urban History of Science. In: Dierig, S., Lachmund, J., and Mendelsohn, A. (eds.), *Science and the City (Osiris 18)*. Chicago: The University of Chicago Press, 1–19.

Diogo, M. P., Gavroglu, K., and Simões, A. (2016). STEP Forum Special Issue. *Technology and Culture* 57:4, 926–997.

Eisenstadt, S. N. (2000). Multiple Modernities. *Daedalus* 109:1, 1–29.

Ewen, S. (2008). Transnational Municipalism in a Europe of Second Cities: Rebuilding Birmingham with Municipal Networks. In: Saunier, P.-Y., and Ewen, S. (eds.), *Another Global City: Historical Explorations into the Transnational Municipal Moment, 1850–2000*. New York: Palgrave Macmillan, 101–117.

Ewen, S. (2015). *What is Urban History?* Cambridge: Polity Press.

Exenberger, A., and Strobl, P. (2013). Introduction. In: Exenberger, A., Mokhiber, J., Strobl, P., and Bischof, G. (eds.), *Globalization and the City: Two Connected Phenomena in Past and Present*. Innsbruck: Innsbruck University Press, 11–21.

Forgan, S. (2005). Building the Museum: Knowledge, Conflict, and the Power of Place. *Isis* 96:4, 572–585.

Gantner, E., and Hein-Kircher, H. (2017). "Emerging Cities": Knowledge and Urbanization in Europe's Borderlands 1880–1945: Introduction. *Journal of Urban History* 43:4, 1–12.

Gantner, E., Hein-Kircher, H., and Hochadel, O. (2018). Backward and Peripheral? Emerging Cities in Eastern Europe (Special Issue). *Zeitschrift für Ostmitteleuropa-Forschung* 67:4.

Gantner, E., Hein-Kircher, H., and Hochadel, O. (eds.) (2020). *Interurban Knowledge Exchange in Southern and Eastern Europe, 1870–1950.* London: Routledge.

Girón, Á. (2017). De redes informales e historias cruzadas: Barcelona-Buenos Aires y la gestión libertaria del conocimiento científico hacia 1900. In: Girón, Á., Hochadel, O., and Vallejo, G. (eds.), *Saberes transatlánticos. Barcelona y Buenos Aires: conexiones, confluencias, comparaciones (1850–1940).* Aranjuez: Doce Calles, 159–186.

Girón, Á. (2019). *Darwinismo, política del significado y redes transnacionales anarquistas: lecturas de la lucha por la existencia entre Londres, París, Barcelona y Buenos Aires (ca. 1879- ca.1910).* VIII Coloquio internacional sobre Darwinismo en Europa y América. El Evolucionismo en Canarias, Ruiz, R., Puig-Samper, M. Á., and Sarmiento, M. (eds.). Madrid: UNAM—Edición Doce Calles, 327–344.

Girón, Á., Hochadel, O., and Vallejo, G. (eds.) (2017). *Saberes transatlánticos. Barcelona y Buenos Aires: conexiones, confluencias, comparaciones (1850–1940).* Aranjuez: Doce Calles.

Global Urban History (2016a). "World History Needs More Urban Mess": A Conversation with Carl H. Nightingale. Published 8 August 2016, https://globalurbanhistory.com/2016/08/08/rare-is-the-messy-stuff-of-urban-history-a-conversation-with-carl-h-nightingale/. Last accessed 18 June 2020.

Global Urban History (2016b). "Urban and Global History Have Been Converging": A Conversation with Shane Ewen. Published 18 August 2016, https://globalurbanhistory.com/2016/08/18/urban-and-global-history-have-been-converging-a-conversation-with-shane-ewen/. Last accessed 18 June 2020.

Goebel, M. (2015). *Anti-Imperial Metropolis: Interwar Paris and the Seeds of Third World Nationalism.* Cambridge: Cambridge University Press.

Golinski, J. V. (1998). *Making Natural Knowledge: Constructivism and the History of Science.* Cambridge and New York: Cambridge University Press.

Griffiths, J. (2009). Were there Municipal Networks in the British World c.1890–1939? *The Journal of Imperial and Commonwealth History* 37:4, 575–597.

Guarneri, J. (2017). *Newsprint Metropolis: City Papers and the Making of Modern Americans.* Chicago: The University of Chicago Press.

Gunzburger Makas, E., and Damljanovic Conley, T. (eds.) (2010a). *Capital Cities in the Aftermath of Empires: Planning in Central and Southeastern Europe.* London: Routledge.

Gunzburger Makas, E., and Damljanovic Conley, T. (2010b). Introduction: Shaping Central and Southeastern European Capital Cities in the Age of Nationalism. In: Gunzburger Makas, E., and Damljanovic Conley, T. (eds.), *Capital Cities in the Aftermath of Empires: Planning in Central and Southeastern Europe.* London: Routledge, 1–28.

Hietala, M. (1987). *Services and Urbanization at the Turn of the Century: The Diffusion of Innovations.* Helsinki: Finnish Historical Society.

Hochadel, O., and Nieto-Galan, A. (eds.) (2016a). *Barcelona: An Urban History of Science and Modernity, 1888–1929*. London and New York: Routledge.

Hochadel, O., and Nieto-Galan, A. (2016b). How to Write an Urban History of STM on the "Periphery". *Technology and Culture* 56:4, 978–988.

Hochadel, O., and Nieto-Galan, A. (eds.) (2018). *Urban Histories of Science: Making Knowledge in the City, 1820–1940*. New York and London: Routledge.

Inkster, I., and Morrell, J. (eds.) (1983). *Metropolis and Province: Science in British Culture, 1780–1850*. London: Hutchinson.

Kargon, R. H. (1978). *Science in Victorian Manchester: Enterprise and Expertise*. Manchester: Manchester University Press.

Kenny, N., and Madgin, R. (eds.) (2015). *Cities beyond Borders: Comparative and Transnational Approaches to Urban History*. Farnham: Ashgate.

Lafi, N. (2008). Mediterranean Connections: The Circulation of Municipal Knowledge and Practices at the Time of the Ottoman Reforms, c.1830–1910. In: Saunier, P.-Y., and Ewen, S. (eds.), *Another Global City: Historical Explorations into the Transnational Municipal Moment, 1850–2000*. New York: Palgrave Macmillan, 35–50.

Levin, M. R., Forgan, S., Hessler, M., Kargon, R. H., and Low, M. (eds.) (2010). *Urban Modernity: Cultural Innovation in the Second Industrial Revolution*. Cambridge, MA and London: MIT Press.

Livingstone, D. N. (2003). *Putting Science in Its Place: Geographies of Scientific Knowledge*. Chicago: The University of Chicago Press.

Livingstone, D. N., and Withers, C. W. J. (eds.) (2011). *Geographies of Nineteenth-Century Science*. Chicago: The University of Chicago Press.

Marsden, B. (2018). Institutionalizing the 'Metropolis of Mechanics': Philosophical Engineering in the City of Glasgow c. 1820—c. 1875. In: Hochadel, O., and Nieto-Galan, A. (eds.), *Urban Histories of Science*. London: Routledge, 37–58.

Middell, M., and Naumann, K. (2010). Global History and the Spatial Turn. *Journal of Global History* 5:1, 149–170.

Møller Jørgensen, C. (2017). Nineteenth-Century Transnational Urban History. *Urban History* 44:3, 544–563.

Morrell, J. (1985). Wissenschaft in Worstedopolis: Public Science in Bradford, 1800–1850. *British Journal for the History of Science* 18:1, 1–23.

Nye, M. J. (1986). *Science in the Provinces: Scientific Communities and Provincial Leadership in France, 1860–1930*. Berkeley, Los Angeles, and London: University of California Press.

O'Sullivan, T. (2019). *Geographies of City Science: Urban Lives and Origin Debates in Late Victorian Dublin*. Pittsburgh: University of Pittsburgh Press.

Phillips, D. (2012). *Acolytes of Nature: Defining Natural Science in Germany, 1770–1850*. Chicago: The University of Chicago Press.

Purdue, O. (ed.) (2013). *Belfast: The Emerging City, 1850–1914*. Dublin and Portland, OR: Irish Academic Press.

Rentetzi, M., and Flevaris, S. (2018). Envisioning a New European Metropolis; Designing the Athens Observatory (1842). In: Hochadel, O., and Nieto-Galan, A. (eds.), *Urban Histories of Science*. London: Routledge, 16–36.

Roberts, L. (2009). Situating Science in Global History: Local Exchanges and Networks of Circulation. *Itinerario* 33, 9–30.

Sakurai, A. (2013). *Science and Societies in Frankfurt am Main*. London: Pickering & Chatto.

Sastre, J., and Valentines, J. (2016). Technological Fun: The Politics and Geographies of Amusement Parks. In: Hochadel, O., and Nieto-Galan, A. (eds.), *Barcelona: An Urban History of Science and Modernity, 1888–1929*. London and New York: Routledge, 92–112.

Saunier, P.-Y., and Ewen, S. (eds.) (2008). *Another Global City: Historical Explorations into the Transnational Municipal Moment, 1850–2000*. New York: Palgrave Macmillan.

Schneer, J. (1999). *London 1900: The Imperial Metropolis*. New Haven: Yale University Press.

Secord, J. A. (2004). Knowledge in Transit. *Isis* 95:4, 654–672.

Shapin, S. (1988). The House of Experiment in Seventeenth-Century England. *Isis* 79:3, 373–404.

Stühlinger, H. R. (2017). Jean-Charles-Adolphe Alphand: Les Promenades de Paris, 1867–1873. In: Lampugnani, V. M., Albrecht, K., Bihlmaier, H., and Zurfluh, L. (eds.), *Manuale zum Städtebau. Die Systematisierung des Wissens von der Stadt 1870–1950*. Berlin: DOM Publishers, 30–49.

Terán, O. (2000). *Vida intelectual en el Buenos Aires fin de siglo (1880–1910). Derivas de la cultura científica*. Buenos Aires: Fondo de Cultura Económica.

Umbach, M. (2005). A Tale of Second Cities: Autonomy, Culture, and the Law in Hamburg and Barcelona in the Late Nineteenth Century. *The American Historical Review* 110:3, 659–692.

Vallejo, G. (2017). Barcelona en la cultura científica argentina del cambio del siglo XIX al XX. De Sarmiento a Fors. In: Girón, Á., Hochadel, O., and Vallejo, G. (eds.), *Saberes transatlánticos. Barcelona y Buenos Aires: conexiones, confluencias, comparaciones (1850–1940)*. Aranjuez: Doce Calles, 104–131.

Valls, L. (2019). Natura cívica: ciència, territori i ciutat al parc de la Ciutadella de Barcelona a principis de segle XX. PhD dissertation, Universitat Autònoma de Barcelona.

Werner, M., and Zimmermann, B. (2006). Beyond Comparison: *Histoire Croisée* and the Challenge of Reflexivity. *History and Theory* 45:1, 30–50.

Withers, C. W. J., and Livingstone, D. N. (2011). Thinking Geographically about Nineteenth-Century Science. In: Livingstone, D. N., and Withers, C. W. J. (eds.), *Geographies of Nineteenth-Century Science*. Chicago: The University of Chicago Press, 1–19.

Wood, N. (2006). Urban Self-Identification in East Central Europe Before the Great War: The Case of Cracow. *East Central Europe/ECE* 33/1-2: 11–31.

Wood, N. (2010). *Becoming Metropolitan: Urban Selfhood and the Making of Modern Cracow*. DeKalb: Northern Illinois University Press.

Zarzoso, A., and Martínez-Vidal, À. (2016). Laboratory Medicine and Surgical Enterprise in the Medical Landscape of the Eixample District. In: Hochadel, O., and Nieto-Galan, A. (eds.), *Barcelona: An Urban History of Science and Modernity, 1888–1929*. London: Routledge, 69–91.

Part Two

Focus on Vienna 1

Technical and Science-Based Infrastructures, 1850–1875

4 The Beginnings of the "City Machine"

Infrastructure Expansion and International Technology Transfer in Vienna, 1850–1875

Sándor Békési

Introduction

"The world we inhabit is a technical world," the philosopher Max Bense stated around the middle of the twentieth century (cited in Ropohl 2009, 15). This insight has become even more valid in the meantime, not least with regard to large cities. From its very beginnings, the city in particular as a special place had been geared toward relieving its inhabitants of certain work processes on the one hand and of the inconvenience of nature on the other (Siebel 2015, 109ff). In order to guarantee this, extensive construction measures and technical infrastructures were (and are) necessary, which in the second half of the nineteenth century reached a quantitatively and qualitatively new stage of development in European and North American cities. At that time, the technical-urban environment that is still familiar to us today was essentially created, which is currently to be further developed with the help of digitalization in the direction of the 'Smart City'.

In this process of modernization, urban technologies not only play a role as an accompanying phenomenon of urbanization, they also have a constitutive meaning, for the modern industrialized city is not possible without a multitude of technical infrastructures. The emerging metropolis of the nineteenth century required innovative systems of supply, disposal, mobility, and communication. On the other hand, such infrastructure facilities were developed and manufactured especially for growing urban needs. At the same time, urban technologies stimulated economic growth and the physical transformation of the city. There was a close interaction between the growth of urban technology and the growth of the city (Radkau 1994, 77; Hughes 1999, 212). Thus, "technical incorporation", for example in the form of tram lines leading into the suburbs, often preceded political incorporation. The tramway let the overflowing urban space 'shrink' again, and at the same time, physical mobility became increasingly substitutable by electronic means of communication (Schott 1999, 6f.). In any case, technologies increasingly began to have a significant influence on urban living conditions and people's everyday lives (Lees and Lees 2007, 137).

The period around 1850 can generally be regarded as a watershed or as a moment of compression in the context of urbanization, modernity, and technological development (Tarr and Dupuy 1988; Levin 2010, 11). In the second half of the nineteenth century, both the first industrial networking of the city and also the era of the first "global integration" took place. This included the construction of supply and disposal systems (water, sewage, gas, electricity) as well as communication and local transport systems (telegraph, tramway, telephone), but also the regulation and reconstruction of entire districts or waters (Schott 1999, 2–8; Rodogno et al. 2015, 3). This period can be further divided into a first phase from around 1850 to 1875, still marked by the so-called first industrial revolution, including steam engines, gas lighting and telegraph, and a second period toward the end of the nineteenth century, which was already part of a second industrial revolution, using electricity or the telephone. In the period up to the end of the 1870s, the course was often set for the assumption of tasks of public services by the municipal administrations. In addition, international or transnational networks of experts intensified and institutionalized and, last but not least, found expression in specialist and world exhibitions (Reulecke 1985, 62; Hughes 1999, 211; Lees and Lees 2007, 138; Rodogno et al. 2015, 7).

As a result of this interaction between technologies and urbanization, a culturally restructured and artificially designed urban space emerged more and more in the course of modernity, which could also be called "Technotop" (Ropohl 2009, 15). But this analogy is not intended to imply a technological determinism, for it is not technology alone that makes the city. Rather, the engineered and networked metropolis was the result of a combination of engineering sciences, architecture and urban planning, and business activity, as well as city administration and politics.

Science and technology played a central role in these changes, as did urban elites, their ideology of progress, and the new institutions they created. All of these urban techniques required the development and application of scientific and technical knowledge. It is well known that the city has long been a major location for most scientific institutions and knowledge production. Less well known, however, is the extent to which science, through infrastructure and technology, influenced city life, structure, and appearance. We are dealing mainly with technical or medical sciences, applied by hygienists, engineers, architects, and urban planners. As Carroll Purcell writes: "If the modern city was not so obviously the child of science, it was evident to all that it existed only through the midwifery of engineering" (Purcell 1995, 131, see also 214f.). But this process involved more than the simple application of scientific methods to urban problems. The question arises to what extent and how there was an interaction between the scientific laboratory and the city (Dierig et al. 2003, 1–3, 8, 15f.).

At the same time, the provision of rapidly growing cities with new infrastructures and urban technologies during the so-called Founders' Period (*Gründerzeit*) from the middle of the nineteenth century onwards was only possible, at least in Central Europe, through the cross-border international exchange of knowledge and technology transfer. The new city was thus from the standpoint of technologies intraurban as well as interurban, based on physical networking on the one hand and on the other hand on informational networking between industrial locations as well as trade and science centers in Europe and beyond, especially in North America.

1. Physical and Informational Networking

The comprehensive technical interconnection of the city is regarded as one of the most striking and momentous phenomena of urbanization in classical modernity (Schott 1999, 2). The common feature of these new infrastructures was that they not only created isolated or individual linear structures in urban space, but also increasingly formed branched networks within the city. The network became a material and symbolic construct at the same time, with the help of which the hygienic, traffic, and supply problems of the metropolis were to be solved or represented. Thomas Hughes's term 'large technical systems' also applies in this context: network character, large-scale expansion, and capital intensity are among the most important characteristics of such systems. Large technical systems extended beyond the technical artifacts (such as electric dynamos) themselves, but were characterized by the interaction of economic, political, and technical-scientific factors. Classic examples are the railway system or urban technology such as street lighting (Hård and Misa 2008, 14; Joerges and Braun 1994, 21–23; Kleinschmidt 2007, 65).

The different infrastructural systems of the city are not to be seen in isolation from each other; rather, they interact with one another and only contribute to the growth of the city through their interaction. Just in this way, the "city machine" and an overall socio-technical system were created (Schott 1999, 7). The installation of such large technical systems simultaneously led to a certain path dependency. According to Hughes's phase model, the first phase of invention and development is followed, second, by actual transfer of technologies from one region to another as soon as they are ready for the market, and finally, third, by a phase of growth, accompanied by a tendency toward diminished innovation in established large-scale systems (Zumbrägel 2015, 95–98).

As Joachim Radkau notes, "Networking also happens continuously without a grand plan," which means that large networks of urban technologies had already begun to develop at a time when there was hardly any

institutionalized urban planning (Radkau 1994, 55). The technical transformation of the city from the middle of the nineteenth century onwards
is usually described as a reaction to growing problems of hygiene, energy
supply, and increased inner-city distances (see also Klemun, Chapter 5
in this volume). With rapid growth and the perception of the emerging metropolis as a crisis, numerous measures for city-wide supply of
water and energy, for sewage disposal as well as for the installation of
communication and local transport systems, were indeed justified at that
time (cf. Schott 2014, 278). However, new infrastructures were often
created by private initiative and on the basis of capitalist exploitation
interests, which the then liberal-minded city administrations tried to influence in a more or less directive way. When it came to protection against
epidemics, for example, strengthening municipalities themselves did not
always act out of humane considerations or for the benefit of all citizens.
Implicitly, on the other hand, the aim at that time was also to promote
the development of industrial capitalist society with the help of technical
infrastructure—that is, to improve the conditions for commercial and
industrial production as well as the conditions for the workforce (Melinz
and Zimmermann 1996, 140f.). But city governments' own fiscal interests
could also be a motive for the introduction of new urban technologies,
because municipal utilities often generated considerable revenues in the
form of fees.

However, the reconstruction and accelerated technical networking of
the city were not limited to specific projects and demands, but were also
driven and guided by more or less explicit planning paradigms. The predominant model of circulation, which guided the reorganization of urban
space, aimed to avoid stagnation of air, water, and waste by assuring an
orderly and unhindered 'flow'. This way of thinking was based not least
on the idea, widespread at the time, that 'miasms' of bad air originat-
ing from fermentation and rotting processes caused diseases, but also
turned out to be in line with the later bacteriological approach. This
figure of thought emphasizing flows included not only hygiene, but also
traffic mobility. Around the middle of the nineteenth century, people first
thought about the city in such terms, with the help of allegories such as
"organism" or "net" (Heidenreich 2004, 175). The image of a 'machine',
on the other hand, was not increasingly transferred to the city until later,
namely after 1945 in the context of criticisms of postwar modernism. But
the functional approach was probably effective much earlier as a guiding
principle for thought and action (Knie and Marz 1997, 97–108; Hård and
Misa 2008, 14ff.). Accordingly, the goal was to coordinate the individual
infrastructure areas of the city (machine) more and more efficiently, and
above all to assure that they functioned smoothly.

The use and importance of machines and technology have always
been, and still are, ambivalent. While on the one hand they are liberat-
ing and servicable, on the other they have a structuring and disciplining

effect. The incipient production of the modern city meant a gradual substitution of self-sufficiency and independent transport by external supply via technical networks, above all at the level of energy and water supply, but with restrictions also in transport. These new technologies thus required far-reaching adaptations on the part of urban populations, at the same time promoting their integration into everyday life and increasing control of their behavior (Schott 1999, 733; Siebel 2015, 111). Last but not least, internationally widespread urban technologies had a fundamentally normative effect on the cityscape and urban structures: In many respects, the engineered, industrialized metropolis was similar everywhere. At the same time, the inherent tension between homogenization and differentiation remained intact in the production and reshaping of European cities by modern technology. Thomas Hughes coined the term "technological style" to describe the shaping and adaptability of technology as a function of culture (cited in König 1999, 218f.; see also Hård and Misa 2008, 1, 12).

In this way, technical infrastructure as a whole became one of the primary factors that could be used to demonstrate the 'rise' and 'progress' of a city and thus its participation in the dynamic development of the industrial age (Wilding 1999, 250f.). This perception and a pronounced optimism about progress based on a belief in technology convey the following quote from a widely circulated contemporary journal: "Today, the institutions of a metropolis include not only lighting cables, water pipes, local railways, electric telegraphs, but also pneumatic mail, or as one could express it, 'letter lines'. . . . If only everyone has his letter pipe device, like his electric telegraph, his water pipe and his gas outflow apparatus, then we are one big step closer to the 'perfect' world" (*Über Land und Meer* 1876, 774). Progress became materialized, so to speak, in the form of the modernized city. In this process the engineer, the inventor, or the bold entrepreneur came to the fore: they were, as it were, the heroes of the bourgeois-liberal age (Levin 2010, 11; Sieferle 1984, 146).

As already mentioned, this development would hardly have been possible without knowledge and technology transmission. This took two basic forms: vertical transfer from research to application, and horizontal transfer of technologies from one region to another (Troitzsch 1983, 177; Braun 1992, 17). Although the present chapter addresses both processes, its focus is mainly on the latter, above all on the international transfer of technical knowledge, artifacts, services, and persons for the purpose of urban infrastructure development. In the case of Vienna—in contrast to the relationship between colonial powers and their colonies—the transfer between industrialized countries with the same economic order and approximately the same technological level was involved. There were, of course, differences amongst countries and regions in the degree of industrialization, and thus the aspect of a catch-up modernization or the closing of a 'technological gap' must be considered. During the period under study, England in particular, although it was gradually losing its dominant position as an industrialized country, was still regarded as the world's technological and industrial center, but so were France and the United States (Braun 1992, 19f; König and Weber 2003, 113–121). From an Austrian perspective, however, Germany was also an important exporter of technology. In the course of time, as will be shown next, transfer developed from a one-way street to a multilateral exchange process. However, a one-to-one transfer of technical solutions was not always possible, because solutions that had been effective elsewhere had to be adapted and modified to new internal, and external, economic and cultural structures (Jeremy 1991, 2f.).

Technology transfer took place through different channels and took different forms (here from the point of view of the recipient country): (1) import of machines and equipment; (2) employment of foreign engineers; (3) agreements with foreign companies; and (4) transfer of technical-scientific knowledge (Pacey 1990, 155). In order to achieve this, appropriate practices and activities such as study trips, competitions, publications, or congresses were necessary (Levin 2010, 9f.). Business enterprises played an important role in this process not only as distributors, but also as intermediaries between science and practical applications. But the determinants of technical diffusion were manifold and included not only economic, but also cultural and social factors (Kenwood and Lougheed 1982, 12; Jeremy 1991, 2). The technological innovation process ranged from scientific discovery (for example in the field of hygiene, communication, or gas production) to patenting and concrete, problem-oriented applications, in our case in an urban context, for example a machine for transporting water, a pressure pipe for sewers, or a telegraph (Krabbe 1985, 79).

This gradually emerging urban expertise can hardly be reduced to individual academic disciplines, but often consisted of a heterogeneous patchwork of overlapping knowledge from different disciplines, blurring the boundaries between academic and administrative knowledge cultures (Dierig et al. 2003, 7). A typology of the experts involved in urban technology at the time could distinguish between the individualistic scientist or technician, who was largely independent of established institutions, the administrative official involved and trained in the field, and the self-employed entrepreneurial engineer (also as a project developer) or even "intellectual entrepreneur" (Forgan 2010, 79; Dienel 1998, 13). We could add to this list the politically active scientist, such as Eduard Suess in Vienna, who knew how to combine his university or academic role with a commitment to public urban issues or even with a political function at the community level (see Marianne Klemun, Chapter 5 in this volume). What is explained previously using London as an example could also apply to Vienna during this period: "The diversity of urban scientific elites in this period should be emphasized, with no sharp division made between academic elites, civil scientists, and men . . . who bridged different organizations and networks. Such men mixed freely across various organizations, moving between public and private enterprise and working alongside professional engineers, industrialists, academic research scientists, liverymen from the ancient City guilds, and consultants to the LCC [London County Council, SB]" (Forgan 2010, 113).

In the following section, Vienna will be used as an example to illustrate how infrastructural expansion and technological networking changed a city and what role the international transfer of knowledge and goods played in this process. Essentially, these were processes and phenomena that became effective and could be observed in large European and North American cities, but also in small and medium-sized towns during the second half of the nineteenth century. However, certain inequalities in infrastructure development were not only observed between cities, but also within them (Reulecke 1985, 56–60; Zimmermann 1996, 31, 172). In this respect, Vienna is probably a representative and at the same time significant example of an international development, while at the same time the question of site-specific factors and characteristics arises.

2. Case Study Vienna

After the middle of the nineteenth century, Vienna also experienced a massive surge in development on its way to becoming an industrialized metropolis. Hardly any other major city in Europe underwent such radical reconstruction in such a short period of time. A large number of infrastructure projects gradually transformed the city into a "Technotop" (Ropohl) and a culturally strongly overstretched space interspersed with "large technical systems" (Hughes). Essential elements or foundation stones for this development were laid in the period from about 1850 to 1875. This does not mean that nothing had happened before in the infrastructural development of the city, or that there had not been many precursor systems (water pipes, sewers, etc.) (see Mayer 1999, 311f; Meißl 2005, 151, 157). But in the phase discussed here there was a massive compression and intensification of relevant measures and projects. The work of this early founding period created the basic technical, health, and organizational conditions required for a large city. This took place in the sense of technical development on the basis of industrially produced artifacts and technical-scientific knowledge, but not yet in the context of large-scale industrial production; Vienna initially remained above all a residential and consumer city in this period. Accordingly, a lower demand for technical infrastructure was present, apart from the construction of the representative Ringstrasse. It was not until the mid-1870s that industrial capitalist production conditions were able to assert themselves to a greater extent (Banik-Schweitzer 1978, 91; Seliger and Ucakar 1985, 434f., 566). The 1873 World Exposition in Vienna, a gigantic economic and cultural exhibition with international significance, gave special expression to this phase of economic and social optimism (more on this later on in the chapter).

Local conditions and triggers for intensified urban development in Vienna in this period were above all rapid population increase, the physical expansion of the city in 1850, municipal autonomy granted in 1862, and a backlog of demand for industrialization and economic prosperity from the mid-1860s onward. The population of Vienna within today's borders practically doubled from around 550,000 in 1850 to 1.16 million around 1880. The former inner suburbs were incorporated and the city's area increased more than tenfold. A liberal municipal constitution and local self-administration were established and became in a remarkable way the central actor in the creation of urban infrastructure, and so turned Vienna into an "active city" (Melinz and Zimmermann 1996, 141f.; Maderthaner 2006, 191–204). In the years 1848 to 1876, as the city's budget grew rapidly, expenditures for so-called technical infrastructure (including water pipes and sewage works) were always significantly higher than those for social infrastructure (education or health) (Pichler 1991, 781, 786). In this phase, Vienna also opted to award concessions to private companies for most infrastructure projects, with the exception of the waterworks and the sewage system; the comprehensive municipalization of infrastructure companies and their transfer to city management began only later, as was the case in most comparable cities.

At the same time, Vienna's development was embedded in overarching international trends and dependent on certain framework conditions. These included, for example, the predominant ideology and political power of liberalism, expansion of technical and scientific activity between about 1840 and 1880 (Krabbe 1985, 15f.; König and Weber 2003, 14), accelerated urbanization, and, last but not least, innovations in hygiene, urban technology, and transportation, which took place in many places at the time (Lees and Lees 2007, 131–146; Kleinschmidt 2007, 13f.). The suppliers of technical systems were often foreign companies that tried to open up new markets here through public tenders or their own offers.

Who were the actual constructors and the designers of the modern metropolis? In this context, three important actors or groups of actors should be highlighted: entrepreneurs, municipalities, and engineers (Schott 2014, 281–284). Around the middle of the nineteenth century, engineers were not only responsible for technical matters, but in many cases also for urban planning. The major urban expansion and conversion plans of that time in Europe were not yet designed by architects, a major exception being Vienna itself, with the Ringstrasse competition of 1858. It is also remarkable that in Vienna the urban and technical design of the city is not associated with the name of a single personality, as in Hobrecht's Berlin, Cerdà's Barcelona, or Lindley's Hamburg. Instead, an influential political-scientific duo can be identified here, whose work influenced several major infrastructure projects and thus the development of the early founding city: Cajetan Felder and Eduard Suess. Both had, among other things, decisive influence on the project planning of the mountain source water system, called the High Spring Water Source Pipeline (*Hochquellen-Wasserleitung*), and the regulation of the Danube near Vienna; their alliance thus offers a remarkable example of the connection between science and administration. Felder was mayor of Vienna from 1868 to 1878, while Suess was university professor of geology and an internationally recognized scholar who also served as a member of the Vienna city council (Békési 2014, 95; on Suess see Marianne Klemun, Chapter 5 in this volume).

With his exploration of the geological conditions of the city, Suess established a scientific basis for large-scale projects in the traffic sector, in the construction of water pipes, or for the construction of the central cemetery. But other renowned scientists were also active at the same time in the political life of the city, by serving in the city council or in other ways: among them were the architects Ludwig Förster, Carl Hasenauer, and Friedrich Schmidt, or the art historian Rudolf Eitelberger. The reform of higher technical education after 1848 (Mikoletzky 1995b) proved to be important beyond Vienna, as theoretical and scientific knowledge became more and more relevant to modern production processes and large-scale construction projects. For example, progress in structural analysis was an essential prerequisite in civil engineering, and Georg Rebhann rendered outstanding service in this field at the Imperial and Royal Polytechnic Institute (builder of the Aspernbrücke, 1863–64; Winkler, 1873, 26). With Emil Winkler he founded the Vienna School of Bridge Construction. In 1878 a municipal testing institute for building materials was set up, which was later institutionalized as a department of the building authority. Rudolf Niernsee, who held this position from 1865 to 1877, acquired his polytechnical knowledge by private studies, while his successor Hieronymus Arnberger completed formal polytechnical training (Csendes 2003, 8–11; Tillmann 1935, 21–24, 42–44). An important forum for scientific exchange and international technology transfer was offered by the Austrian Society of Engineers, founded in 1848, and by its journal (Mikoletzky 1995a, 115).

3. New Urban Infrastructures

The following is a brief overview of the most important urban infrastructures that were newly built or massively expanded in Vienna between 1850 and 1875. These concerned the areas of energy and water supply, drainage and disposal, transport and telecommunications.

1. Water supply: The first water pipeline in Vienna to supply the entire urban area (including the inner suburbs) was the Kaiser Ferdinand

water pipeline, which was commissioned in the mid-1840s. This drew water from the Danube Canal and directed it with the help of steam engines (using English technology) to higher parts of the city. The rapidly increasing demand for water in the developing city and repeated epidemics brought new impulses in the 1860s. This concerned not only sanitation and the improvement of drinking water quality, but also the provision of sufficient quantities of industrial water for the irrigation of gardens or for street and sewer cleaning. At that time the city took the controversial decision to transport water to Vienna from the high mountains about 80 kilometres to the south in the form of a risky and costly High Spring Water Source Pipeline (*Hochquellen-Wasserleitung*) on its own initiative, although there were also nearby alternatives. The aforementioned Eduard Suess played an important role as a proponent of the project, as a member of the city council and also of the municipal water supply commission, but the Society of Physicians advocated this option for hygienic reasons as well. However, it was probably not only scientific findings and advice, but also the consumption needs of the upper middle classes that helped to achieve the breakthrough (Kortz 1905, 215f.; Seliger and Ucakar 1985, 541, 563; Békési 2014, 96–98). The contractor Antonio Gabrielli from London was commissioned to head construction of the water plant. In addition to several Austrian companies, numerous foreign firms from Germany, England, Belgium, and Liechtenstein were also involved in the project as suppliers of building materials or cast iron pipes (Winkler 1873, 72, 78; Wehdorn 1979, 391). The opening of the Vienna High Spring Water Source Pipeline and the representative high-beam fountain at the Schwarzenbergplatz took place in 1873. The ambitious but still inadequate project became the most expensive infrastructure project of the liberal era in Vienna and contributed significantly to the city's indebtedness (Figure 4.1)

2. Sewer system: With regard to the disposal of waste water, Vienna already possessed extensive older facilities. The city within the bastions was already almost completely canalized in the eighteenth century. The first main or collecting canals were built on the Vienna River on the occasion of the cholera epidemic of 1831/32, and were considerable installations by international standards. In the course of the 1860s, the uniform canalization of urban expansion areas was promoted and the organization of canal clearance was gradually taken over by the local authorities. New sewage projects abroad (London, Frankfurt am Main, Berlin, or Gdansk), the findings of the English River Pollution Commision, and the work of the chemist Max von Pettenkofer in Munich on hygiene served as models and resources for these projects, which treated urban drainage from new scientific points of view (Kortz 1905, 194f.; Seliger and Ucakar

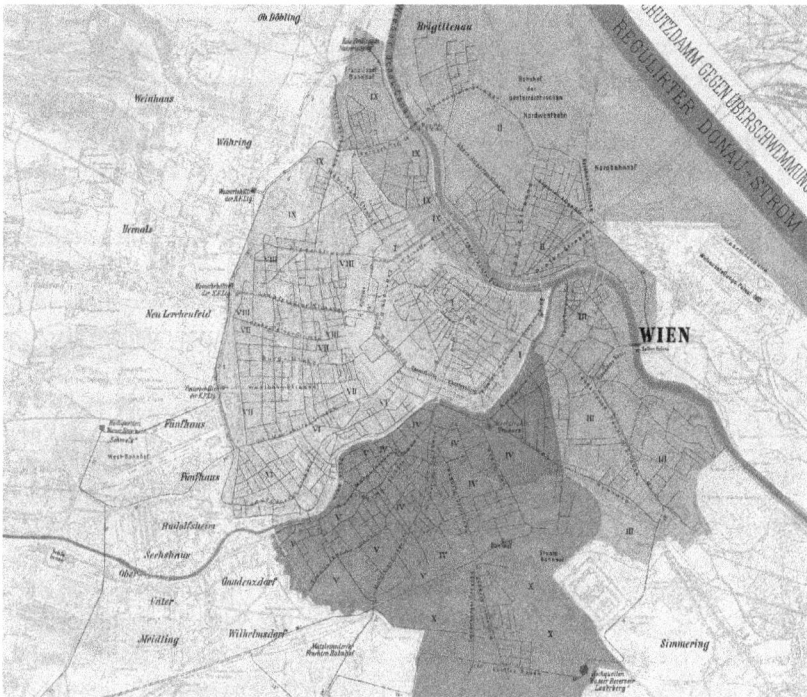

Figure 4.1 Pipeline network (solid lines) and pumping stations of the Vienna High Spring Water Source Pipeline, 1876.

Source: Gemeinde-Verwaltung der Reichshaupt- und Residenzstadt Wien. Wien Museum, inventory no. HMW 8480.

1985, 525; Radkau 1994, 79). A uniform sewer system for the districts of Brigittenau and Favoriten and the then Donaustadt in the second district was already constructed with concrete sewers, in which a company from Germany was also involved or a German patent applied. By 1873, the year of the World Exhibition, around 90 percent of the houses in Vienna were connected to the sewer network. Soon afterwards, the decision was made to maintain the alluvial sewer system; the relatively early turn away from the barrel system was promoted also by the city's favorable morphology and water supply situation (Kortz 1902, 44f.; Wehdorn 1979, 130).

3. Gas supply: The first installations of this type were used primarily for lighting buildings and streets and were built in Vienna from 1828 onwards. The English company Imperial Continental Gas-Association (ICGA) acquired the Fünfhaus gasworks in 1842 and gradually purchased all the small gasworks in Vienna. In 1855 the

Österreichische Gasbeleuchtungs-Aktiengesellschaft (ÖGA) was founded to counterbalance the expansion of the ICGA. However, the private English gas company remained dominant in Vienna for decades and was the subject of fierce discussions about an approach in this service sector that was hardly oriented toward the common good or determined by foreign firms. Ultimately, this was not only about technology but also about capital transfer, especially since ICGA also used dumping prices to put existing local gas producers under pressure (Meiß 2005, 157). In 1862 there were already more than 6,000 lighting fixtures in the Vienna city area, and on May 1, 1865, the gas lighting of the Vienna Ringstrasse was ceremoniously put into operation. At this time, six gas works in the city are already supplying energy. In 1875, the municipality of Vienna concluded a new lighting contract with ICGA, which remained valid until 1899 (Kortz 1905, 241; Ruck and Fell 2009, 96–111, 119).

4. Telegraphy: The first modern form of information technology and telecommunications was not only interurban, but also included electric telegraphy within the city itself. The first line was built in Austria in 1846 between Vienna and Brno according to the system of Alexander Bain (Scotland), with public use permitted from 1850. The introduction of the telegraph was also closely linked to the railway system and the economy as a whole: Andreas Baumgartner (physicist, manufacturer, and politician) initially headed the establishment of the electrical telegraph network, and soon after became director of the Emperor Ferdinand Nordbahn and a full member of the Imperial and Royal Academy of Sciences. In 1865 the Viennese telegraph manufacturer and engineer C. A. Mayrhofer made the first proposal for the establishment of a local private telegraph supply in Vienna according to the English motto: "time is money." The project was realized a few years later without him under state control and in a slightly modified form (Herzog and Pensold 2010, 117, 126f.; Wehdorn 1979, 413). The local telegraph network in Vienna and the surrounding area already comprised 38 stations by around 1870. In addition to the gas lighting and the tracks of the tramway, telegraph poles as elements of physical infrastructure and networks were now clearly visible in the cityscape (Meißl 2005, 156).

5. Pneumatic tube system: The regular pneumatic tube system in Vienna was put into operation in 1875 in response to the increasing capacity problems of postal and telegram transport and the economy's need for faster communication. The 'pneumatic telegraph' transported practically 100 telegrams within the city at a cost that would otherwise have been required for the electrical transmission of a single telegram. At the same time its operation required a complex system of steam engines, air pumps, and air reservoirs (see Figure 4.2).

The individual stations were interconnected by an underground network of pipes in which pneumatic trains of interconnected metal

Figure 4.2 The pneumatic post in Vienna, 1876.
Source: *Über Land und Meer* 18:39, 774. Wien Museum, inventory no. HMW 238317.

cans were moved at high speed with the aid ofan artificially generated air pressure difference in front of and behind the train. The
Vienna Tube Mail (*Wiener Rohrpost*) was originally developed
according to the French model (poligonal system) by a technician
and industrialist from Austria, Franz Felbinger, and manufactured
in Vienna. By the turn of the century, the pneumatic tube network
had grown to around 50 kilometres in length and 49 stations. After
the success of the pneumatic tube system in Vienna, Felbinger was
soon called to Berlin to set up a similar plant there, and also exported
his system to other German cities(Kortz 1905, 153f.; Herzog and
Pensold 2010, 126f.; Wehdorn 1979, 415).

6. Railway stations: As stated previously, the railway is regarded as
    the prototype of modern technology and large-scale technical systems
    (Radkau 1994, 68). Vienna's railway stations, the first of which
    were built in 1838, were the starting and end points of this physical
    communication network. Between 1859 and 1873, this first genera-
    tion of railway stations was rebuilt or massively expanded, and
    three new locations were added. Within one and a half decades, six
    large terminus stations were built and operated in Vienna by private
    railway companies from Austria and abroad. In the construction of
    the stations, the achievements of the engineers were closely linked

to those of the architects. Only a few of the designers of the six new major railway stations in Vienna were anchored in local tradition. Most of them originally came from Württemberg, then one of the centers of railway construction in Europe. This was also true of the architect of the Ostbahnhof, Carl Schumann, who had already become a Ringstrasse architect in Vienna before he completed it in 1870. The building's French style elements can also be regarded as a tribute to the French investors in the railway company concerned (Kassal-Mikula and Haiko 2006, 87, 91f.).

7. Streetcar lines (tramway): Apart from the large railway stations, which in Vienna were mostly built on the outskirts and initially isolated from each other, it was the horse tramway that brought rail lines into the city and into urban space. As a result, a new physical means of communication and a faster, more efficient form of transport were created, by further developing the horse omnibus. There were several private parties from Austria and abroad interested in the construction of such a facility in Vienna following the American and Western European model. The Austrian engineer Gustav von Dreyhausen was initially unsuccessful with his project of a 'Viennese rail' and a 'Viennese wagon', and left his system to a Geneva construction company, which was willing and able to raise the necessary funds for a test track in 1865. A few years later the Wiener Tramway-Gesellschaft also received a concession from the municipality, and in 1872 the Neue Wiener Tramway-Gesellschaft was added as a competitor. By the world exhibition year 1873, a basic network with ring lines and radial routes to the suburbs had been created (Winkler 1873, 32f.). In contrast to the water supply system and similar to the gas supply system, the liberal city administration left the operation of inner-city traffic to private companies, while retaining individual control options.

8. Mountain railways: On the occasion of the 1873 World Exhibition, private operators built three so-called mountain railways in Vienna and the surrounding area. With their help, attempts were made to make the hills outside the city, called 'mountain sections', which had been arduous to reach until then, more convenient for tourists and locals alike, and to make the eating establishments and hotel businesses built there more easily accessible. Different technologies were used here: The Leopoldsberg was reached by a cable car operated by the Austrian Mountain Railway Company, which at the time was one of the best of its kind in the world, alongside the Ofner cable car (now in Budapest) or the one in Lyon, which may have served as models here. Only the ropes came from Germany. The rack railway on the neighboring Kahlenberg was again built on the model of the Rigi railway in Switzerland, with important components obtained from France (Winkler 1874, 7f.). On the other hand, the cableway to the Sophienalpe was designed by the Viennese

manufacturer Georg Sigl, who also patented it. In an energy-efficient circulation system, so-called fiaker carriages for four persons each were attached to a continuously moving towing device as required (Kos and Gleis 2014, 516–518).

9. Bridges: In connection with the regulation of the Danube in 1869–1875 and the expansion of railway traffic, five large stone river bridges were built in Vienna within a few years—in addition to numerous bridges across the Vienna River and the Danube Canal—including two road bridges and three railway bridges. They formed an important link not only in the transport network, but also for other technical systems, in which they served as carriers of gas and telegraph lines. Of these, the North Railway Bridge was practically the only one that came into being without foreign participation. The East Railway Bridge was built, among others, by the French company Castor & Comp., which shortly before had completed the Suez Canal and a little later carried out the construction work for the Danube regulation (Figure 4.3). For the Nordwestbahn bridge and the road bridge to Floridsdorf, a German company supplied the iron superstructure, while the iron construction for the Reichsbrücke bridge came from another French construction company (Winkler 1873, 28–31). In addition to the large railway stations, important new signal points in the urban space were created in the form of bridges with their impressive engineering structures (Kos and Gleis 2014, 348–351).

Figure 4.3 Building the Stadlau bridge in Vienna, 1870.

Source: *Allgemeine Bauzeitung*, 1870, Blatt 75. Wien Museum, inventory no. HMW 58635/1.

4.   The Radical Transformation of the City

A special feature of the Viennese example is the fact that the infrastructural expansion was accompanied by large-scale physical transformations of the city, including the Ringstrasse construction, the Danube regulation, and the World Exposition. Above all the first two examples represent a production of urban space according to the planning motto "the straight path is the best" (Förster), which in a broad sense can be considered as being itself technical infrastructure, or whose production required such infrastructure to a great extent. These massively redesigned urban spaces, the dimensions of which were unparalleled in Vienna as a whole up to that time, became essential features of the new metropolis.

Ringstrasse zone: This boulevard, which—after removal of the old city fortifications—was essentially completed by 1865, was known at the time as one of Europe's largest urban planning projects. The international competition of 1858 for the basic plan of this inner city expansion is considered to be the first of its kind in Europe; the cost was borne exclusively by the state (Stühlinger 2015, 10). Among the participants whose designs were included in the final plans was the Prussian garden architect Peter Joseph Lenné. For the construction of the road itself as well as the entire Ringstrasse zone, it was necessary to level the earth uniformly, which required extensive earth movements; some sections were accomplished by removal of land, others by filling (Psenner 2018, 7f.). Finally, the Ringstrasse was created as a complex "multiway boulevard", consisting not only of side lanes and avenues but also of riding paths and—as an innovation—a parallel, subordinate load road (Békési 2015, 74). In addition, an infrastructure of water pipes, sewer lines, gas supply, and tramway tracks was gradually added. The technical design of this new urban space also included ingenious and extensive ventilation and heating systems, which connected the monumental buildings along the street to the infrastructure via huge underground shafts (Wehdorn 1979, 106, 226). It should be noted that the renewal of the road system through regulation and paving was also part of infrastructure development and improvement of mobility conditions in other parts of the city at that time (Seliger and Ucakar 1985, 534–540, 565).

Regulation of the Danube: The major regulation of the Danube near Vienna from 1869 to 1875 arose out of a complex set of interests, including flood protection, trade policy, and urban expansion. This project was carried out in equal parts by the governments of the city, the state, and the province of Lower Austria. In our context it is first of all important to emphasize that this measure included a radical straightening of the river and a massive redesign of the urban space, which until then had been without precedent in the neighborhood of a large city. This willingness to take risks was rooted in a general enthusiasm for progress and belief in technology that were predominant at the time. The regulation of the

Danube near Vienna represents a special stage in the history of industrialization: Measures were taken on a large scale to change the landscape on the basis of fossil fuels, in this case the excavation of a new, approximately 13-kilometre-long river bed with the aid of steam engines (Figure 4.4) in order to facilitate the passage of steam ships and railways—based again on the use of fossil energy (Békési 2016).

The engineers James Abernethy from London and Georg Sexauer from Karlsruhe played a major role in the planning of the Danube regulation. The French company Castor, Couvreux & Hersent was commissioned to carry out the construction work. The technology employed had also been successfully used in the construction of the Suez Canal immediately before. While the technicians and engineers came mainly from France, the simple workers were recruited from Slovakia, Bohemia, Silesia, and Italy, as was the case with other larger construction projects in this country (Kos and Gleis 2014, 321–329).

World Exhibition: This major event of 1873 was organized by private initiators and operators and was only hesitantly supported by the municipality of Vienna. Naturally, the show itself included an increased transfer of technology and a demonstration of the latest international industrial culture

Figure 4.4 Floating dredger with locomotive used during the regulation of the Danube near Vienna, 1870–1876.

Source: Photograph by Hermann Voigtländer. Wien Museum, inventory no. HMW 48865/2.

(Felber and Camilleri 2004). The focus here is on the construction of the exhibition site and the central building, the Rotunda. At 233 hectares, the total area of the Prater site surpassed fivefold that of the 1867 World Exhibition in Paris; for the first time, an extensive, multipart pavilion system was used. Several French and German companies were also involved in the construction of the site. The site was technically highly developed, including a railway network (essential for transporting the exhibits), around 1,500 lighting fixtures, electrically illuminated fountains, several waterworks with an extensive network of lines for the necessary utility and drinking water, sewers, and rinsing water for the first water closets in Vienna (Kos and Gleis 2014, 359, 364–370). Thus, an artificial city was created within the city, which in infrastructural terms often anticipated the modern metropolis. The Rotunda became a gigantic symbol of technical progress and a landmark of the city. At that time it was the largest freely spanned interior in the world, and it could hold around 20,000 visitors. The building was originally designed by the English engineer John Scott-Russel, and the iron construction was carried out by a Duisburg company (Figure 4.5). In this way, formal inventiveness and technical-industrial know-how from abroad were combined (Kos and Gleis 2014, 380).

Figure 4.5 Josef Langl: Vienna at the time of the World Exposition, 1873, detail with exhibition park and buildings. Oil on canvas.

Source: Wien Museum, inventory no. HMW 16864.

Summary

The year 1873 brought both a climax and a turning point in the development of Vienna's technical infrastructure. The stock market crash and the economic crisis that followed marked the temporary end of several major projects in the city, including the construction of a local railway or light rail network and a central railway station, as well as the regulation of the Vienna river in connection with the construction of a new boulevard (Maderthaner 2006, 204f.; Békési 2014, 104f.). From this point of view, the 'city machine' of Vienna around 1875 was only partially realized in comparison to its original potential and to other European metropoles. However, essential foundations had been laid and some of the major steps taken on the way to a modern metropolis in this period from 1850 to 1875. All this was only possible through a diversified transfer of knowledge and technology as well as the exchange of goods and services, characterized by a high degree of internationalization and networking. As Rodogno, Struck, and Vogel put it generally, "scientific and technical experts became agents of the emergence of a transnational or, in some cases, supranational consciousness among European élites" (Rodogno et al. 2015, 2). This phenomenon can be studied especially profitably in metropoles of the *Gründerzeit*, because it was during this time that large cities increasingly became centers of technological knowledge and innovations that had begun in the early nineteenth century.

Although many aspects of the Viennese case are locally specific, this example illustrates general trends in technology transfer and urban development in Europe and North America at that time. Here, too, as elsewhere, various elements and characteristics of international metropolitan development and discourse of the time, such as claims of prestige, competition among cities, forced modernization, and optimism about progress, became effective. But after 1873 this dynamism collapsed; Vienna's claim to becoming a "cosmopolitan city" (Winkler 1873, IX) failed at first. If one understands metropolises as comparatively large cities with international significance in economic, political, and cultural terms and at the same time as centers of innovation and creative milieus (Hall 1998, 18–22, 281), then Vienna as a whole can probably be counted as such a metropolis only around 1900, although the term appears earlier in local media (for the latter point see Klemun, Chapter 5 in this volume). Nonetheless, in the field of urban technology the city proved to be an active recipient and user of foreign know-how in the third quarter of the nineteenth century. Austrian participation or innovation largely appears to have failed to come into play at first, but Vienna provided innovative and independent services of international standing, above all in three large-scale enterprises—the Ringstrasse project, the Danube regulation, and the World Exposition.

Bibliography

Anonymous (1876). Die pneumatische Post in Wien. *Ueber Land und Meer. Allgemeine Illustrirte Zeitung* 18:39, 774.

Banik-Schweitzer, R. (1978). Liberale Kommunalpolitik in Bereichen der technischen Infrastruktur Wiens. In: idem. (ed.), *Wien in der liberalen Ära* (Forschungen und Beiträge zur Wiener Stadtgeschichte, Bd. 1). Vienna: Verein für Geschichte der Stadt Wien, 91–119.

Békési, S. (2014). Auf dem Weg zur Stadtmaschine? Zur Infrastrukturentwicklung Wiens in der frühen Gründerzeit. In: Kos, W., and Gleis, R. (eds.), *Experiment Metropole—1873: Wien und die Weltausstellung*. Vienna: Czernin-Verlag, 94–105.

Békési, S. (2015). Hauptverkehrsader, Promenade—und Barriere. Zur Verkehrsfunktion der frühen Ringstraße. In: Nierhaus, A. (ed.), *Der Ring. Pionierjahre einer Prachtstraße*. Vienna [St. Pölten, Salzburg, Wien]: Residenz-Verlag, 72–76.

Békési, S. (2016). Zwischen Hochwasserschutz und Stadtpolitik. Zur Entstehung des Donau-Durchstiches bei Wien 1870–1875. In: Tamáska, M., and Szabó, Cs. (eds.), *Donau-Stadt-Landschaften: Budapest—Wien*. Berlin: Lit-Verlag, 229–244.

Braun, H.-J. (1992). Technologietransfer: Theoretische Ansätze und historische Beispiele. In: Pauer, E. (ed.), *Technologietransfer Deutschland-Japan von 1850 bis zur Gegenwart*. Munich: Iudicium-Verlag, 16–47.

Csendes, P. (2003). Stadt und Technik. Wissenschaftlicher Fortschritt und urbane Entwicklung. In: Archiv der Hauptstadt Budapest, Wiener Stadt- und Landesarchiv (eds.), *Budapest und Wien. Technischer Fortschritt und urbaner Aufschwung im 19. Jahrhundert*. Budapest and Vienna: Eigenverlag, 5–14.

Dienel, H.-L. (1998). Zweckoptimismus und-pessimismus der Ingenieure um 1900. In: idem. (ed.), *Der Optimismus der Ingenieure. Triumph der Technik in der Krise der Moderne um 1900*. Stuttgart: Steiner, 9–24.

Dierig, S., Lachmund, J., and Mendelsohn, A. (2003). Introduction: Toward an Urban History of Science. In: idem. (eds.), *Science and the City* (*Osiris* 18). Chicago: University of Chicago Press, 1–19.

Felber, U., and Camilleri, C. (eds.). *Welt ausstellen: Schauplatz Wien 1873*. Vienna: Technisches Museum Wien, 2004.

Forgan, S. (2010). From Modern Babylon to White City: Science, Technology and Urban Change in London 1870–1914. In: Levin, M. R. et al. (eds.), *Urban Modernity: Cultural Innovation in the Second Industrial Revolution*. Cambridge, MA and London: MIT Press, 75–132.

Hall, P. (1998). *Cities in Civilization*. New York: Pantheon Books.

Hård, M., and Misa, T. J. (eds.) (2008). *Urban Machinery: Inside Modern European Cities*. Cambridge and London: MIT Press.

Heidenreich, E. (2004). *Fließräume. Die Vernetzung von Natur, Raum und Gesellschaft seit dem 19. Jahrhundert*. Frankfurt am Main and New York: Campus Verlag.

Herzog, M., and Pensold, W. (2010). Die Anfänge des modernen Kommunikations- und Medienwesens. In: Rumpler, H. (ed.), *Die Habsburgermonarchie 1848–1918: IX. Soziale Strukturen, 1/1: Lebens- und Arbeitswelten in der Industriellen Revolution*. Vienna: Verlag der Österreichischen Akademie der Wissenschaften, 109–157.

Hughes, T. P. (1999). Berlin: Technological Metropolis. In: Goodman, D. (ed.), *The European Cities & Technology Reader: Industrial to Post-Industrial City.* London and New York: Routledge, 211–220.

Jeremy, D. J. (1991). Introduction. In: idem. (ed.), *International Technology Transfer: Europe, Japan and the USA, 1700–1914.* Aldershot: Edward Elgar, 1–5.

Joerges, B., and Braun, I. (1994). Große technische Systeme – erzählt, gedeutet, modelliert. In: idem. (eds.), *Technik ohne Grenzen.* Frankfurt am Main: Suhrkamp, 7–49.

Kassal-Mikula, R., and Haiko, P. (2006). Vom "Arsenalstil" zur "Wiener Renaissance". Wiens gründerzeitliche Bahnhöfe in baukünstlerischer Sicht. In: Kos, W., and Dinhobl, G. (eds.), *Großer Bahnhof. Wien und die weite Welt.* Vienna: Czernin Verlag, 86–101.

Kenwood, A. G., and Lougheed, A. L. (1982). *Technological Diffusion and Industrialisation Before 1914.* London et al.: Croom Helm.

Kleinschmidt, Chr. (2007). *Technik und Wirtschaft im 19. und 20. Jahrhundert* (Enzyklopädie deutscher Geschichte, 79). Munich: Oldenbourg.

Knie, A., and Marz, L. (1997). *Die Stadtmaschine. Zu einer Raumlast der organisierten Moderne* (Discussion Paper FS-II 97–108). Berlin: Wissenschaftszentrum Berlin.

König, W. (1999). *Künstler und Strichezieher. Konstruktions- und Technikulturen im deutschen, britischen, amerikanischen und französischen Maschinenbau zwischen 1850 und 1930.* Frankfurt am Main: Suhrkamp.

König, W., and Weber, W. (2003). *Netzwerke. Stahl und Strom 1840 bis 1914* (Propyläen-Technikgeschichte, 4). Berlin: Propyläen Verlag (unveränd. Neuausg. von 1990).

Kortz, P. (1902). Entwässerung. In: Weyl, Th. (ed.), *Die Assanierung von Wien* (Die Assanierung der Städte in Einzeldarstellungen, Bd. 1, H. 2). Leipzig: W. Engelmann, 41–65.

Kortz, P. (1905). *Wien am Anfang des XX. Jahrhunderts. Ein Führer in technischer und künstlerischer Richtung* (Bd. 1). Vienna: Gerlach & Wiedling.

Kos, W., and Gleis, R. (eds.) (2014). *Experiment Metropole—1873: Wien und die Weltausstellung.* Vienna: Czernin-Verlag.

Krabbe, W. R. (1985). *Kommunalpolitik und Industrialisierung: die Entfaltung der städtischen Leistungsverwaltung im 19. und frühen 20. Jahrhundert* (Schriften des Deutschen Instituts für Urbanistik, 74). Stuttgart et al.: Verlag W. Kohlhammer/Deutscher Gemeindeverlag.

Lees, A., and Lees, L. H. (2007). *Cities and the Making of Modern Europe, 1750–1914.* Cambridge: Cambridge University Press.

Levin, M. R. (2010). Dynamic Triad: City, Exposition, and Museum in Industrial Society. In: Levin, M. R., et al. (eds.), *Urban Modernity: Cultural Innovation in the Second Industrial Revolution.* Cambridge, MA and London: MIT Press, 1–12.

Maderthaner, W. (2006). Von der Zeit um1860 bis zum Jahr1945. In: Csendes, P., and Opll, F. (eds.), *Wien. Geschichte einer Stadt, Bd. 3: Von 1790 bis zur Gegenwart.* Vienna, Cologne, and Weimar: Böhlau Verlag, 175–544.

Mayer, W. (1999). Technische Infrastruktur. In: Csendes, P., and Opll, F. (eds.), *Die Stadt Wien* (Österreichisches Städtebuch, 7). Vienna: Verlag der Österreichischen Akademie der Wissenschaften, 311–328.

Meißl, G. (2005). Vernetzung: Wie die Stadt zur Maschine wurde. In: Brunner, K., and Schneider, P. (eds.), *Umwelt Stadt. Geschichte des Natur- und Lebensraumes Wien* (Wiener Umweltstudien, Bd. 1). Vienna: Böhlau, 150–161.

Melinz, G., and Zimmermann, S. (1996). Die Aktive Stadt. Kommunale Politik zur Gestaltung städtischer Lebensbedingungen in Budapest, Prag und Wien (1867–1914). In: idem. (eds.), *Urbanisierung, Kommunalpolitik, gesellschaftliche Konflikte (1867–1918): Wien—Prag—Budapest.* Vienna: ProMedia, 140–176.

Mikoletzky, J. (1995a). Der "österreichische Techniker": Standespolitik und nationale Identität österreichischer Ingenieure, 1850–1950. In: Plitzner, K. (ed.), *Technik, Politik, Identität.* Stuttgart: GNT-Verlag, 111–123.

Mikoletzky, J. (1995b). Vom Polytechnischen Institut zur Technischen Hochschule. Die Reform des technischen Studiums in Wien, 1850–1875. *Mitteilungen der Österreichischen Gesellschaft für Wissenschaftsgeschichte* 15, 79–100.

Pacey, A. (1990). *Technology in World Civilization: A Thousand-Year History.* Oxford: Basil Blackwell.

Pichler, E. (1991). Kommunalwirtschaft: Die quantitativen Dimensionen der öffentlichen Hand. In: Chaloupek, G., Eigner, P., and Wagner, M. (eds.), *Wien. Wirtschaftsgeschichte 1740–1938, Teil 2: Dienstleistungen.* Vienna and Munich: Jugend & Volk, 757–846.

Psenner, A. (2018). 'Wiener Null': Levelling the City of Vienna. In: *Urban Research & Practice*, Published online: 11 Sep 2018.

Purcell, C. (1995). *The Machine in America: A Social History of Technology.* Baltimore: John Hopkins University Press.

Radkau, J. (1994). Zum ewigen Wachstum verdammt? Jugend und Alter großer technischer Systeme. In: Joerges, B., and Braun, I. (eds.), *Technik ohne Grenzen.* Frankfurt am Main: Suhrkamp, 50–106.

Reulecke, J. (1985). *Geschichte der Urbanisierung in Deutschland.* Frankfurt am Main: Suhrkamp.

Rodogno, D., Struck, B., and Vogel, J. (2015). Introduction. In: idem. (eds.), *Shaping the Transnational Sphere: Experts, Networks and Issues from the 1840s to the 1930s.* New York and Oxford: Berghahn, 1–20.

Ropohl, G. (2009). *Allgemeine Technologie. Eine Systemtheorie der Technik* (3rd ed.). Karlsruhe: Universitäts-Verlag.

Ruck, H., and Fell, Ch. (2009). *Gas: Energie für Wien im Wandel der Zeit*, Bd. 1. Wien: Wien-Energie-Gasnetz-GmbH.

Schott, D. (1999). *Die Vernetzung der Stadt. Kommunale Energiepolitik, öffentlicher Nahverkehr und die "Produktion" der modernen Stadt. Darmstadt—Mannheim—Mainz 1880–1918.* Darmstadt: Wissenschaftliche Buchgesellschaft.

Schott, D. (2014). *Europäische Urbanisierung (1000–2000): Eine umwelthistorische Einführung.* Stuttgart: UTB, Cologne [u.a.]: Böhlau.

Seliger, M., and Ucakar, K. (1985). *Wien, politische Geschichte 1740–1934. Entwicklung und Bestimmungskräfte großstädtischer Politik. Band 1: 1740–1895* (Geschichte der Stadt Wien, Bd. 1). Vienna: Verlag Jugend und Volk.

Siebel, W. (2015). *Die Kultur der Stadt.* Berlin: Suhrkamp.

Sieferle, R. P. (1984). *Fortschrittsfeinde? Opposition gegen Technik und Industrie von der Romantik bis zur Gegenwart.* Munich: Beck.

Stühlinger, H. R. (2015). *Der Wettbewerb zur Wiener Ringstraße: Entstehung, Projekte, Auswirkungen.* Basel: Birkhäuser.

Tarr, J. A., and Dupuy, G. (eds.) (1988). *Technology and the Rise of the Networked City in Europe and America.* Philadelphia: Temple University Press.

Tillmann, R. (1935). *Festschrift zur Hundertjahrfeier des Wiener Stadtbauamtes.* Vienna: Deutscher Verlag für Jugend und Volk.

Troitzsch, U. (1983). Technologietransfer im 19. und 20. Jahrhundert. *Technikgeschichte* 50:3, 177–180.

Wehdorn, M. (1979). *Die Bautechnik der Wiener Ringstraße mit einem Katalog technischer Bauten und Anlagen in der Ringstraßenzone* (Die Wiener Ringstraße, Bild einer Epoche, Bd. 11). Wiesbaden: Verlag Franz Steiner.

Wilding, P. (1999). Technik und Urbanität: Der Ausbau der technischen Infrastruktur als Leitmotiv städtischer Modernisierung in Wien und Graz um 1900 (Studien zur Moderne 4). In: Uhl, H. (ed.), *Kultur—Urbanität—Moderne. Differenzierungen der Moderne in Zentraleuropa um 1900.* Vienna: Passagen-Verlag, 243–286.

Winkler, E. (ed.) (1873). *Technischer Führer durch Wien.* Vienna: Lehmann & Wentzel.

Winkler, E. (ed.) (1874). *Technischer Führer durch Wien. Ergänzungen.* Vienna: Lehmann & Wentzel.

Zimmermann, C. (1996). *Die Zeit der Metropolen. Urbanisierung und Großstadtentwicklung.* Frankfurt am Main: Fischer.

Zumbrägel, Chr. (2015). Dreißig Jahre danach: Thomas P. Hughes' *Networks of Power* als Leitkonzept der Stadt- und Technikgeschichte. *Informationen zur modernen Stadtgeschichte* 1, 93–98.

5 Metropolitan Geology and Metropolitan Collections

Turning Vienna into Stones in the Nineteenth Century

Marianne Klemun

Introduction

From the mid-nineteenth century, Vienna as the capital of the Habsburg Monarchy transformed itself into an industrial city and experienced a huge population increase (Weigl 2000; Baltzarek 1980). While town fortifications in other major European capitals had been taken down as early as in the seventeenth century, the medieval city of Vienna had remained enclosed within fortification walls that elsewhere had lost their importance long before. A strip of land several hundred metres wide outside the walls, the so-called Glacis, surrounded the city. In 1857, the emperor decided that Vienna should be relieved of its medieval walls, which he deemed an awkward urban relic. Instead, the construction of the Ringstrasse, a boulevard with ambitious imperial architecture, was initiated (Wagner-Rieger 1972–1981). The new road was intended to connect the city with its surrounding villages and suburbs (Mayer 1972), and the infrastructure of an expanding urban space was to be developed. This marked the first step of an urban expansion that gave the city a new and modern metropolitan face. It took 30 years to complete all of the monumental buildings that had been planned.

This study departs from the observation that geology and not only artistic, architectural, and technological innovations connected the emergence of Vienna as a metropolis and the new image of the city. Geologists such as Franz von Hauer (1822–1899), Eduard Suess (1831–1914), and others recognized the indispensable need to identify the geological layers underneath the city of Vienna in connection with its waterscapes and underground water systems. This happened already in the late 1850s and early 1860s, when the decision in favor of the great Ringstrasse project was made, to provide scientific arguments in public debates about improving water quality and supply for the rapidly growing population. As a consequence of this research, the course of the meandering Danube River was regulated (1870–1875), a new water pipeline (1873) was built that connected Vienna with the Alps, ensuring a steady supply of good-quality water for the city, and a new central cemetery was established outside the city (for details and further analysis, see Chapter 4 by Békési, in this volume).

The claim advanced here is that visionary geologists, rather than architects or technicians alone, set the course for a new urban infrastructure in Vienna. They recognized that the water resources of the city were limited and that poor water quality was responsible for diseases. Eduard Suess in particular put his finger on the city's crucial problems. In his extraordinary book, *Der Boden der Stadt Wien nach seiner Bildungsweise, Beschaffenheit und seinen Beziehungen zum bürgerlichen Leben. Eine geologische Studie (The Soil of Vienna according to its Formation, Nature and its Relations to the Lives of its Citizens. A Geological Study)* (Suess 1862a), he discussed the city's dependence on natural water resources and its geographical location. Suess combined the findings of various fields of knowledge concerning urban space and linked them with the analysis of geological strata and the social situation of the city.

The Berlin-based scholar Dorothee Brantz has introduced the term "thick space" (Brantz et al. 2012) into academic research on metropolises, in contrast to the mostly normative uses of the notion "density" in urban studies, which had become a "potentially calculable metaphor" (Roskamm 2011) rather than a coherent concept. In my analysis of Eduard Suess's metropolitan geology, I will take up Brantz's concept and refer not only to the metropolis as a specific space, but also to different international research approaches that converged in a 'thick space', an agglomeration of knowledge in and about the metropolis, including medicine and hygiene, epidemiology, hydraulic engineering, statistics, architecture, demography, and finally stratigraphy. As I will show, Eduard Suess combined all of these approaches and created a 'thick space' of knowledge in the metropolis Vienna.

The theoretical concept of "urban metabolism" (Wolman 1965) seems particularly useful for the analysis of Eduard Suess's geological research. It was introduced by Abel Wolman in 1965, and has since been widely used for different approaches in urban studies (Kennedy et al. 2007); it is particularly useful for combining historical changes and environmental or ecological perspectives (Gingrich et al. 2012). The concept has been employed to facilitate the description of the flows of materials and energy within cities, and has provided researchers with a metaphorical framework to analyze the interactions of natural and human systems at a specific place. According to Kennedy urban metabolism is "the sum total of the technical and socio-economic processes that occur in cities, resulting in growth, production of energy and elimination of waste" (Kennedy et al. 2007, 43). I would argue that supply and discharge, input and output—the key aspects of a metabolism—are conditions that geologists already took into consideration in the 1850s, when they analyzed underground water flows of rivers and outflows or exchanges between different sedimentary materials in and beneath the city of Vienna. Thus, this chapter presents an integrated socio-cultural and scientific perspective on the interrelation between Vienna's urban metabolism and geological knowledge.

Decades before metropolitan studies were explicitly established and increasing attention was paid to social and cultural aspects of urban life, which happened around 1900 (Simmel 1903), geologists claimed European capital cities such as London, Paris, and Vienna as new fields of research for themselves. I would like to call this new research interest *metropolitan geology*, even though this term was not used by contemporaries. It was not before the 1860s that the term 'metropolis' appeared in Austria in newspapers in popular contexts, referring either to a dominant center or the convergence between industries, traffic, science, cultural events, and good taste realized ideally in New York, London, or Paris. As one writer put it, "The metropolis is the centre of attraction for all creation and inspiration; only what is successful here can influence a lifestyle" (Anonymous 1866). The fact that the city defined what would become or influence a lifestyle was seen as characteristic of a metropolis by contemporaries. Geological considerations of Vienna and its future prospects were increasingly based on or triggered by social and structural developments that were essential features of a metropolis, such as a highly diverse and growing population that coexisted in extreme spatial density. This encounter of geology with the social and cultural perception or awareness of metropolises takes center stage throughout this chapter.

Innovative projects such as the new pipeline that supplied Vienna with fresh water from the Alps (1873) and the regulation of the Danube (1870–1875) are well recognized today, and are still seen as potential advantages for healthy living conditions in Vienna. However, the history of science has hitherto overlooked the fact that these significant advantages were already preceded by Eduard Suess's fundamental theoretical research on the city's geology in 1862. This seminal work therefore deserves closer attention (see Section 3).

To explain Suess's creative contribution, which I propose to define as *metropolitan geology*, it is necessary to examine the context of doing geology in the Habsburg Monarchy and the specific culture of this field of science in Vienna. The first section of this chapter therefore provides a brief overview of the development of geology in the Habsburg Empire and how *metropolitan geology* fit into this context. As a second pillar of *metropolitan geology*, I will focus on specific collections dedicated to the city's materiality that I would like to call *metropolitan collections*. In 1865, the public image of Vienna as a metropolis was grounded in historicist architecture and design. This perspective was supported by and coincided with the establishment of a new type of geological collection of useful stones, which first and foremost demonstrated the materiality of the new and monumental architecture of the Ringstrasse buildings. These new types of collections illustrate the importance of geology for nineteenth-century architecture, and for those features that were considered to be metropolitan, thus embodying urban space. In this way, metropolitan collections and metropolitan geology became relational categories that are both material and abstract.

1. Geology in Vienna: From the Empire to the Metropolis

To answer the question of how geology responded to metropolitan challenges and even adopted central metropolitan topics, it is useful to take a closer look at the unique culture of geology that emerged as a science in the Austrian Empire and especially in Vienna (Klemun 2017). Austria, in this context, refers to the conglomerate of different states under the Habsburg crown that existed until 1918. As in many other European states, the new discipline of geology developed in Austria between the eighteenth and the nineteenth centuries from mineralogy, a field of traditional natural history that served as an umbrella science and was dominated by descriptive instead of theoretical approaches (Rudwick 1996). In contrast to other European states (Rudwick 1963), however, learned societies (Knell 2000) such as the Geological Society of London (founded in 1807) did not exist in Austria. Nonetheless, Austrian geology soon achieved international recognition. Two key aspects shaped the development of the new science and the establishment of public institutions: mining industries outside the city and natural history collections within the capital.

Mining was fundamental to the development of geology in three respects: (1) Many intellectuals and non-experts who worked in the administration of the mining industry were rather enthusiastic about mineralogy and found numerous fields of mineralogical study in the mining facilities alongside their work. (2) From the eighteenth century, the mining school (later academy) in Banská Štiavnica (today Slovakia) trained civil servants to work in public service. Propelled by private initiatives during the 1840s, especially those established by the dedicated Wilhelm von Haidinger (Klemun 2020; see also Chapter 9 by Johannes Mattes in this volume), the central state office of mining, the Court Chamber for Coinage and Mining in Vienna (*Hofkammer für Münz- und Bergwesen*), established a mineralogical collection and offered training courses in the field called "geognosy" (the German term for geology) for civil servants employed at mining facilities throughout the empire. As part of this institution, a collection which specialized in mining (*Montanistisches Museum*) was established in Vienna in order to achieve a higher standard of knowledge among mining officials. It also aimed to establish a representative rock sample collection from all mines in the monarchy, and to compile a geognostical map of the Habsburg lands (Klemun 2017; Bachl-Hofmann et al. 1999). From 1840, these activities were all centralized in Vienna and formed the foundation of the Imperial and Royal Geological Survey (*k. k. Geologische Reichsanstalt*), founded in 1849. The key predecessor of the Reichsanstalt with respect to both personnel and content (Bachl-Hofmann et al. 1999, Hofmann and Klemun 2012), the Mining Museum (*Montanistische Museum*) was established by Wilhelm von Haidinger and Franz Hauer. These were both renowned scientists, who founded several scholarly circles, shaped the scientific discourse in Vienna, and opened the Geological Survey to young interested scholars such as Eduard Suess.

(3) The close connections between mining and public facilities influenced how earth science was conceptualized. Longer than anywhere else, mineralogy, defined more broadly than the current science, was the official term in Austria for what we call earth sciences today. Well-established in Western European languages from the late eighteenth century, the term and concept of geology as a science based on theory and history instead of description met widespread skepticism in Austria. On the other hand, the notion of geognosy (focusing on stratigraphy), initially coined by Abraham Gottlob Werner (1749–1817), who taught at the famous mining school in Freiberg in Saxony, was regarded as more suitable within the framework of a knowledge culture that privileged what Hofbauer has called "vulgar reasoning" (Hofbauer 2003) that prevailed among local mining professionals until the middle of the nineteenth century (Klemun 2015). Eduard Suess, who represented geology in the second half of the nineteenth century not only in Vienna but internationally, criticized geognosy on the grounds that it focused on the mere description of "formations", which had been common before the Geological Survey was established. As he argued: "By this means the historical dimension is excluded. This is not learning to read but merely learning to recognise the letters" (Sueß [*sic!*] 1916, 113). For Suess as a novice in the field of earth science, it was the Geological Survey and the natural history collection of the Imperial Court that turned geology into a space of negotiation and gave it public recognition (Klemun 2009).

After the failed revolution of 1848, during a period of political reaction and Neo-Absolutism, the state initiated several projects to demonstrate the cultural and political unity of Austria. In these efforts historians, philologists, cartographers, geographers, and geologists helped implement the concept of Austria as a cultural nation (*Kulturnation*, see Feichtinger 2010; Feichtinger 2012). The Geological Survey carried out a survey of all territories of the monarchy, resulting in a unified geological map published in 1867. Thus, research served both a political and a scientific purpose (Klemun 2012). But the development of geology as a discipline was also increasingly influenced and accelerated by industrialization and increased consumption during the nineteenth century. From the outset, the Geological Survey saw itself as an institution for applied science that was based on the highest level of reflective theoretical geological research. In 1850, Franz Haidinger, the director of the Survey, outlined the main purpose of the newly founded institution as follows: "The Imperial Geological Survey has a primarily practical purpose: to support practical work by the application of science and to promote science by the power of practical work" (Haidinger 1850, v, vi). Applied science was not seen as a separate activity but integrated into the practice of the Survey. During the first decade, the Survey's goal was to meet a political objective by compiling a unified geological map. "Unity" referred to the Emperor's watchword *Viribus Unitis* (with united forces), and thus to the reconciliation of national differences.

The foundation of the Imperial and Royal Academy of Sciences in Vienna (*k. k. Akademie der Wissenschaften*) in 1847, the reform of the university system in 1849 and, first and foremost, the establishment of the first chairs for geology after 1862 strengthened the position of geology as an academic and theoretical discipline among other sciences (Tollmann 1963; Schübl 2010). However, during the first decades of the second half of the nineteenth century initiatives that we would now call urban or applied geology were launched mostly by protagonists of the Reichsanstalt and the Natural History Court Museum in Vienna. An initial shift from the sole focus on the geology of the entire empire as the dominant goal to that of the capital was noticeable when the Survey was able to register initial success and was able to envisage the completion of the survey of the whole empire after 1859. To the impact of this shift in focus I now turn.

2. Water and Health as Problems of Industrialized Cities: "Thick" Knowledge and Metropolitan Geology Take Form

Studies on metropolises in general refer to urban phenomena as metropolitanism when several different factors that cause urban changes come together. These factors include a specific new attractiveness of the metropolis not only for its inhabitants but also for scientists, urban migration, a growing diverse population, the concentration of wealth in material resources as well as the perception of the city as a space of progress, political centrality, and dense infrastructure (Reif 2012). Complex dynamics among all of these developments are crucial. According to Brantz they can be seen together as creating "thick space" (Brantz et al. 2012), in analogy to Clifford Geertz's view of ethnographic practice as "thick description". In her analysis of metropolises, Brantz has suggested the concept of "thick space" in contrast to the mostly normative uses of the notion "density" in urban studies. In this section I will illustrate how the previously mentioned factors that were studied by different academic disciplines merged into a conglomerate. My argument is that the interaction between these approaches to integrating science and the metropolis created a kind of 'thick space' of knowledge.

Let us first take a look at these different approaches, their nature and reception in Vienna. The cholera epidemics in 1830 and 1854 led to the medical examination of the capital's water supply in order to improve it. Until 1830, it was exclusively physicians who were concerned about the quality of water and air in major cities. Gradually, geologists also contributed to resolving these issues. In Vienna, Joseph Franz Jacquin, professor of botany and chemistry at the medical faculty of the university, and the geognost Paul Partsch drew attention to the so-called artesian wells (Jacquin and Partsch 1831). These wells were drilled into artesian aquifers, with groundwater beneath the water level trapped between pervious strata above and impervious strata below. When drilled into such aquifers, the pressure forced water to the surface. The French physicist Francois Arago (1786–1853) had explained this phenomenon by referring to the principle of communicating vessels. Jacquin and Partsch, who listed 48 developed artesian wells in their publication, carried out test drillings in Vienna. This required a good knowledge of stratigraphy provided by Partsch, the custodian for mineralogy at the Natural History Court Museum.

In Vienna, physicians such as Franz Ragsky, who worked at the Medical-Surgical Joseph's Academy, addressed the water quality of artesian wells (Ragsky 1848). In addition, several scholars began to compare water resources and quality of metropolises internationally, for example the French physiologist Gabriel Grimaud de Caux, who dedicated numerous publications to this topic from 1839 onward (Grimaud de Caux 1839, 1860). Robert Angus Smith (1817–1884), who lived in Manchester, the first industrialized European city, focused on water and air as two elements that were influenced by industrialization (Smith 1849). Smith later determined the acid content of rainwater and became well known as the first environmental chemist. Civil engineers such as John Frederick Bateman (1810–1889) also became involved in the debates on water for the metropolises. Bateman was the most knowledgeable expert on dam construction, water drainage, and supply systems (Bateman 1855). He built magnificent water supply facilities in major British cities, and also planned sites for Buenos Aires, Naples, Constantinople, and Colombo.

The acquisition of water increasingly became the focus of attention through the international comparison of metropolises. The geologists of the Survey in Vienna followed this development closely in the rapidly growing specialized literature: the diaries of Franz Hauer, the first geologist of the Survey, show that he meticulously excerpted details from the publications mentioned previously (Hauer 1859). He focused on the water volume generated by wells and transported through pipes, raising the issue of a potential water shortage. The new Emperor Ferdinand water pipeline, built in 1835 and put into operation in 1841, was intended to prevent a water shortage due to population growth and increasing industrial water consumption. It supplemented the existing six court and six municipal pipelines.

Following the example of English industrial cities, further measures were taken based on the concept of metabolism, which gradually gained ground. In Manchester, drain water was turned into brick-shaped fertilizer by adding agricultural limestone (Anonymous 1857). Between 1802 and 1855, 30 patent certificates were issued in the north English city Anonymous 1858). Thomas Wicksteed, civil engineer in Old Ford, submitted arguably the most famous one on 24 February 1851 (Anonymous 1851) concerning the technology for producing manure from the sludge of the cities. The concept of turning sewage into agricultural fertilizer (Anonymous 1857) was intended to help correct the no longer functioning metabolism of the city. The sewage was formed into a pasty consistency by adding lime milk and centrifuging the deposited solids of the sludge. In so doing, excrement from the cities could be transported to the periphery in a much more discreet way. Hervé Mangon in Paris modified the technology of processing human waste. This cultural technique was intended both to accelerate the disposal procedure and to enable a clean solution of the sewage problem. It is striking that such measures taken in English cities were given great attention in Austrian journals on agriculture and engineering around 1857/1858. All these publications paved the way for the new topic, which was closely monitored and discussed within the Survey (Hauer 1859) and summarized in Eduard Suess's publication (Suess 1862a). However, Suess placed the issue of human waste on a completely new footing by addressing it through the lens of geology. Therefore, his book will be discussed in greater detail in the following section.

3. Eduard Suess's Geology and His Metropolitan Studies: A Brief Overview

This is neither the place to analyze the dynamic development of geology in and beyond Austria after 1850, nor to discuss in detail the changes in the academic system and the universities there after 1849. However, one person stood out in this context due to his achievements on numerous levels: Eduard Suess (1831–1914) was an autodidact palaeontologist, who had worked and studied at the Natural History Court Museum in Vienna. He applied for a postdoctoral qualification—the so-called habilitation—without holding a university degree and received the first professorship in palaeontology, not just in Vienna but in the world, in 1857. He became an expert on stratigraphy and developed the concept of a global earth system science. Although he was also interested in education (Suess 1862b) and public affairs, he devoted the greater part of his life to exploring the evolution of features of the earth's surface (Hofmann et al. 2014). His numerous excursions in the eastern and western Alps encouraged him to think about the process of mountain building. His scientific works covered almost all fields of geology, and his career included nearly all functions in academic life: he was an ambitious museum custodian and fieldworker, was appointed associate professor of palaeontology in 1857, full professor of geology in 1862, and became vice chancellor of the University of

Vienna, and was also president of the Vienna Academy of Sciences from 1898 to 1911. Politically, he was liberal leaning. As a member of the city council of Vienna from 1863 to 1873, and again from 1882 to 1886, he drew public attention to geology. In 1873, he initiated the construction of a 110-kilometre-long aqueduct reaching from the mountains to the growing capital of the monarchy.

Suess's ideas matured slowly, and it was not until 1875, after a quarter of a century of study, that he authored a synopsis and critique of hitherto prevailing theories about the Alps in his book, *Die Entstehung der Alpen (The Origin of the Alps)* (Suess 1875), which became the theoretical foundation of his later work, *Das Antlitz der Erde (The Face of the Earth)* (Suess 1888, ²1892, ³1900). In many ways he integrated the multifaceted knowledge of his time into his works, as Mott T. Greene (1982, 191) has pointed out:

> He recognized, as did his contemporaries, that his genius was not in the spinning out of conjecture or the originality of ideas, but in the working out of an integrated master theory and in the substantiation or refutation of the theoretical claims of others. The notion of paleocontinents was suggested by the paleontological work of Melchior Neumayr [professor for palaeontology in Vienna from 1878]. The idea of zones of diverse displacement (by faulting) and the differentiation of folded and block-faulted mountains came from G. K. Gilbert. The ideas on subterranean Lava reservoirs and maculae originated with Clarence King and Clarence Dutton. The idea of radical subsidence came from Dana and Elie de Beaumont. . . . Studer had turned Suess's attention to the dislocation of the north foreland of the Alps, Escher to the importance of overfolds. . . . He was deeply indebted to the work of Albert Heim and Ferdinand von Richthofen and was inspired by De la Beche, Cuvier, Elie de Beaumont, and Werner.

Suess also created the concept of the spheres of the earth and introduced the terms "atmosphere", "lithosphere", "hydrosphere", and "biosphere" (Suess 1875, 158 and 159; Hofmann et al. 2014). He had a tremendous impact on geological knowledge about stratigraphy and tectonics in the nineteenth and early twentieth centuries.

The high productivity of Suess's interdisciplinary approach was already shown by his early work on the geology of the city of Vienna in 1862. In this publication he combined the knowledge of several different specialized areas, such as medicine, statistics, and geological structures, with the concept of an "urban metabolism", albeit without using this term.

As stated previously, Abel Wolman was the first to argue that cities have a metabolism. In order for them to function they require large amounts of materials, energy, and resource inputs, which leave the city in the form of wastes. Vienna's population never grew faster than between 1800 and

1918; indeed, it rose from 230,000 to slightly more than 2 million citizens in this period, and its territory gradually expanded to its present size. Until 1850, the city had an area of only around 2.8 square kilometres, which corresponds to today's inner city. In 1898, Vienna covered an area of roughly 178 square kilometres. This rapid growth went hand in hand with an increasing demand for more food, energy, coal, water, and other resources as materials for streets and buildings, accompanied by the extension of the city and its transport infrastructure. Density was the other side of the coin of urban expansion. When the urban metabolism expanded, the amount of wastes and excrements—including corpses—grew boundlessly, and geologists reacted to this transformation. Suess's book, *The Soil of Vienna according to its Formation, Nature and its Relations to the Lives of its Citizens. A Geological Study* (Suess 1862a), mentioned previously, was the first that focused on the interrelations of urban metabolism from the perspective of geological analysis (Figure 5.1).

In this publication Suess anticipated topics that would become seminal within the field of metropolitan studies in later years, for instance social and cultural condensation and centralization. However, he discussed these issues based on geological features. Although Vienna's dramatic population growth in the last years of the nineteenth century was not yet foreseeable in 1862, Suess discussed the increase of migrants against the backdrop of the given physical space of the city of Vienna through a geological lens. In this, he was inspired by Charles Darwin's concept of "struggle for life or existence" (Darwin 1859) and Thomas Robert Malthus's *Principle of Population* (Malthus 1826). In Suess's view, the geological approach was the best way to describe the cultural situation of the city and to understand the historical development and the future conditions of this unique place. Suess was enthusiastic about Darwin's work *On the Origin of Species* (Darwin 1859), and was one of the first defenders of its author in Vienna. However, from a palaeontologist's point of view he was skeptical about Darwin's theory (Klemun 2018). In spite of this, he encouraged the debate on Darwin in his discussion groups and adopted the idea of the struggle for life for his own research on the underground spaces of Vienna.

Previously, geologists had been mostly interested in tertiary strata and fossil recognition, particularly in the challenging comparison between the Vienna and the Paris Basins. Although similar fossils had been found in both Vienna and Paris, Suess stressed the difference between these basins due to a steep drop in the subsoil of the Vienna Basin. He had both a social and geological interest in the city of Vienna. In other words, Suess saw the future of the city from the perspective of its natural *and* its cultural history. In connecting the topography of the city and the distribution of geological strata with metropolitan awareness, he had no doubt that the natural condition of the soil impacted the social and cultural life of the city. In his view, Vienna's geology as well as its culture made the city

— 181 —

der Siebenbrünner Wiese, wo wir noch unmittelbar hinter der Linien-Kapelle gemengten Schotter notirt hatten. Die tiefsten Aufschlüsse, die bedeutendsten in der ganzen Stadt, sind jene in den noch in Betrieb stehenden Ziegelgruben am Hungelbrunn und Schaumburger Grund. Man sieht hier an den hohen, fast senkrechten Wänden derselben oben eine 1½—2° starke Lage von rothgelbem Belvedere-Schotter [1]), unter diesem 3'—1° lichteren, sandigen Tegel, dann dunkleren Tegel mit einer Bank von zahllosen Cardien, dann viele Klafter hinab blauen Tegel, von der Ferne schön gebändert durch das Auftreten von rostgelb gefärbten Lagen und von weissen Muschelbänken. Zähne von Hipparion gracile und anderen Landsäugthieren, dann Wirbel und Zähne von Fischen, Tannenzapfen und verkohlte Holzfragmente sind zu wiederholten Malen in diesem Tegel angetroffen worden. Hrn. J. Letocha, dessen Eifer in der Aufsammlung ähnlicher Vorkommnisse ich schon einmal zu erwähnen Gelegenheit hatte, ist es kürzlich gelungen, hier ein Panzerstück einer Sumpfschildkröte zu erhalten.

Fig. 20.

Blech. Thurngasse Nr. 958; Belvedere-Schotter auf Tegel.

Die obere Fläche des Tegels und der Belvedere-

[1]) Auf diese beziehen sich Morlot's Bemerkungen; (siehe S 9 und 10; ebendaselbst sind die Angaben von Hauer und Hörnes über die Gruben am Hungelbrunn angeführt.)

Figure 5.1 A page from Eduard Suess's geological study of Vienna (Suess 1862a, 181), showing layers of rock beneath a house, specifically "Belvedere gravel" on *Tegel*, a marly, clay-like rock containing limestone, considered to be useful for bricks.
Source: Public domain.

unique, and its geological conditions were connected with its cultural decline: "The sunken limestone zone . . . , the collapse, has opened a rift in the mountain divide that lets the Danube take its course and determines the cultural-historical mission of our city" (Suess 1862a, 19). He pointed out that the different geological strata underneath the city had been crucial for its history, for instance in 1529, 1683, 1809, and in 1848, that is in times of war, occupation, or revolution. He argued that the terrain had an influence from where the city was bombarded by the Ottoman conquerors in 1529 and 1683 and Napoleon in 1809.

The cholera outbreaks in 1830, 1836, 1849, and 1854/55 prompted Suess to examine the most important aspect of the city's metabolism, the water-flow beneath Vienna. He investigated the course of underground rivers, their history and, based on chronicle sources, possible changes in the course of time as well as old water pipes and fountains. He contacted master well builders and asked them about their experiences. This research on small rivers, old water pipes, and underground networks helped him form an understanding of the complex geological situation of the Vienna Basin, which is surrounded by the foothills of the Alps. He argued that the underground water pipes and old waterways of Vienna followed a special complex pattern, which confirmed his knowledge of the moon-shaped layers of the tectonic syncline and its sediments.

It should be noted that Suess's focus on public health and the way he compiled data were novel at that time. He began this work before Vienna's Statistical Central Commission (*Statistisches Zentralamt*) was established in 1863 and before this institution published annual reports on statistical data covering the entire Viennese population from 1864 onwards (K. K. Statistical Central Commission 1864). This yearbook, which was an important contribution to sanitation discourse, was without precedent. The first chair for hygiene was not established at the medical faculty of the University of Vienna until 1874. With the outbreak of cholera epidemics, physicians such as Eduard Glatter (1864) became increasingly aware of the issues of public hygiene and sanitation. Due to the lack of other medical statistics, Suess himself consulted the yearbook of the Statistical Central Commission to investigate the spatial distribution of cholera in Vienna as a challenge for geology. He scientifically proved that the city's old cemeteries played a major role in triggering metabolic processes. Suess argued that the uncontrolled flow of organic substances underneath the city of Vienna via old and new small rivers was the reason for the disease outbreaks. Thus, the water system which was so vital for the city could also be a source of danger.

The protagonists of the Geological Survey and the Natural History Court Museum provided the scientific expertise to solve this problem, for instance by identifying the most suitable location for the new main cemetery at the periphery of the city, which was intended to replace the various smaller cemeteries within the city center (Stur 1869). According to their recommendation, a cemetery should be built on non-permeable ground to prevent the uncontrolled flow of organic substances. The growing demands for water and for better water quality were key stimuli for innovation. In this respect, Eduard Suess's geological studies influenced the city's projects and investments in particular. As mentioned previously, he secured Vienna's supply with water from the Alps. However, apart from this and his other well-known contributions—such as a new pipeline system that opened in 1874 and the regulation of the Danube—it is important to emphasize that years before these developments Eduard Suess established the new academic field of *metropolitan geology*. Railroads and tunnels (Wolf 1870) were important in the course of industrialization. After 1870, underground construction projects of the expanding city, such as new buildings, railroads, and fountains, offered new possibilities for geologists to gain short-term insights into hitherto unknown rock outcropings (Wolf 1870, 1874) in order to answer purely scientific questions about the geology of Vienna. The geologists of the Survey thus benefited from the urban take off (Karrer 1877). Years later, in 1874, the Viennese municipal authorities followed Suess's scientific recommendation and established the central cemetery (the *Zentralfriedhof*) on more suitable ground, and the existing cemeteries within the city center were gradually abandoned.

Preliminary considerations on infrastructural issues (for instance the geological determination of water reservoirs that culminated in the construction of the famous Vienna High Spring Water Source Pipeline in 1873; see Békési, Chapter 4 in this volume) and the choice of appropriate locations for the establishment of cemeteries were promising research topics for geologists who followed Suess's ideas in the 1870s. They would prove to be useful for the city and helped establish the new and ambitious applied branch of geology apart from its traditional political tasks that were pursued by the Geological Survey. Geologists were even more socially and culturally influential, because they contributed to the image of the metropolis as a unique site of cultural and natural resources and attracted many people due to its healthy environment. After Suess's contribution in 1862, geologists, like other intellectuals, could not resist the attraction of the metropolis. They connected the deep time of the city with the status of the metropolis as a site favored by fate and rich in structural, natural, and cultural resources.

4.   *Metropolitan Collections* in Vienna

The new buildings of the Ringstrasse were still under construction when the World Exhibition opened its doors in Vienna in 1873. Mineral collections took center stage at this fair. It has often been emphasized that world exhibitions displayed—and epitomized—nationalism, imperialism, consumption, and technological rise (Kretschmer 1999). They were showpiece productions of historicism and progress, connecting monumentality, abundance, speed, statistics, sensations, novelties, and topicality both semantically and visually, thus forming a phantasmagoria. Steel, iron, and glass as insignia of the new era gave the exhibition buildings a special significance as metaphors of progress. Under the roofs of industrial halls, decorated with exotic ornaments and supported by lighting effects, machines, commodities, and goods of knowledge of many different kinds were presented. These "sites of pilgrimage to the fetish of the commodity," as Walter Benjamin (1991) called them, offered education by way of entertainment, edification by way of displaying encyclopedic knowledge as well as identity development, thus participating in competition and comparison among the European nation states. They also offered, for the first time in history, a stage for training and propagating a new concept of resource, connecting national uniqueness, availability, and technological feasibility. This aspect has been hitherto neglected by the rich literature on the Vienna World Exhibition and will therefore be examined here in greater detail.

At the Vienna exhibition the outstanding significance of the realm of minerals as a priority field was expressed by the placement of the mineral display at the physical center of the exhibition, so that it acted as a signifier of the foundation that raw materials seemed to have laid for any civilization. Indeed, this reference played a decisive role by all accounts of world's fairs, as most of them started with the mineral resources. Whereas previously, especially in London, the coal industry had been given priority, new aspects played a role in the presentation of Habsburg power. Most centrally in the Industrial Palace, under its dome with a span of 102 meters, as well as in the Eastern Gallery toward Northern Court 14b, the Geological Survey was given the opportunity to present itself and the "utilizable products of the Habsburg Empire's mineral realm"; through its maps and publications this institution "gave splendid evidence of its more than twenty years of activity" (Klemun 2011, 251).

Apart from statistics and overviews, a collection of minerals consisting of 1,600 pieces and structured into three main displays documented the country's material world, authentically presenting ores, fossil fuels, and building materials (Figure 5.2). The collection of 1,000 stone cubes from the entire territory of the monarchy was something particularly metropolitan. It gave evidence of the variety of possibilities of construction works, and immediately corresponded to the start of the great Ringstrasse project, which marked the transformation of the city from a place enclosed within its medieval boundaries into a modern metropolis.

This unique collection of 1,000 cubes of equal size was curated by Heinrich Wolf (1825–1882) and was displayed in the Geological Survey after the 1873 World Exhibition. Today it is held by the former Mauerbach Monastery, and has been used for the training of architects and restorers since 1994. Wolf, who was initially employed at the Survey as a

Figure 5.2 Collection of useful stones presented at the World Exhibition in Vienna 1873, now displayed in the former Carthusian monastery Mauerbach, near Vienna.

Source: Photograph by Marianne Klemun.

servant and proved himself useful as a backpack bearer during field expeditions, worked his way up and became assistant geologist in 1871 and senior geologist in 1873 (Hauer 1882). His example shows that outsiders could make a career in the Survey and create a niche for themselves, as also shown by Wolf himself, who established the construction collection. This special collection was important for solving the problems that the geologists of the Survey encountered: their expertise of the best construction material was required for the renovation of St. Stephen's Cathedral (Hauer 1859).

At the same time, both the Geological Survey and the Natural History Court Museum focused on the collection of decorative and building stones influenced by Suess's discussion of materials for construction in his book from 1862. Felix Karrer (1825–1903), a lawyer and student of Suess, who never got a post at the museum, separated this collection from others in 1878, after the Union Building Society made a donation of 6,000 pieces. The strong connection between geologists and architects, which is confirmed in Franz Hauer's diary, was crucial to this expertise. Today, the collection holds 25,000 pieces and is the largest in Europe. It is striking that the first museum exhibition of this collection stressed the usefulness of the stones as construction material for the streets of Vienna and their origin from all over the monarchy to construct the new image of the unique location of the Ringstrasse in the metropolis. The collection was a special representation of the new city and its architecture, which focused on the stones of the new buildings, organized by the overall structure of the collection, and demonstrated the richness, centrality, and progress of a city. In 1886, when most of the official buildings were completed, Karrer gave a talk on "Monumental Buildings in Vienna and their Construction Materials" (*Monumentalbauten in Wien und ihre Baumaterialien*) at the Scientific Society (*Wissenschaftlichen Club*) (Karrer 1886). Initially, the collection was limited to specimens from the construction site at the Ringstrasse; however, it was gradually expanded to the entire monarchy and beyond.

Conclusion

The new type of geology that emerged in the mid-nineteenth century explicitly included what was seen as being characteristic features of a metropolis: urban migration, a growing diverse population, wealth in material resources, the city as a space of progress, political centrality, and dense infrastructure as well as its particular new attractiveness. We could call this work *applied geology*, but to give it the significance it deserves, and which was intended by contemporary geologists, the term *metropolitan geology* is more suitable, since it emphasizes the contribution of this type of geology to the development of Vienna as a metropolis in transformation.

Eduard Suess was the main protagonist who developed this new academic field. His 1862 publication tapped into several different fields of knowledge that referred to the metropolis. Epidemiology, hygiene, statistics, technology, and hydraulics played a major role in his considerations of urban metabolism and the exchange between different strata, on underground rivers and waterways. By combining these approaches, he created a 'thick space' of knowledge. Suess's book inspired many of his colleagues to pursue these questions further and to address practical issues. It also encouraged the municipal authorities to establish a central cemetery and a main water pipeline.

It is no coincidence that urban expansion and the construction of the Ringstrasse coincided with the establishment of a new type of collection after 1870, which displayed building materials exclusively. The materiality of the gradually expanding, manifest, and dominating representative architecture at the heart of the city was meticulously presented. The objects in these collections referred to certain stages of earth history, to wide-ranging relations, and the most remote privileged mining locations. They served as visiting cards of uniqueness, concentration, and wealth at world exhibitions and linked the Ringstrasse's houses and their architecture—their "shield" (*Panzer*), as Eduard Suess used to call it—to a concept of deep historical order with which they naturalized and historicized the city of Vienna.

Bibliography

Anonymous (1851). Verzeichnis der vom 27. Dec. 1850 bis 24. Februar 1851 in England ertheilten Patente. *Das Polytechnische Journal* 120, 150–155.

Anonymous (1857). *Allgemeine Land- und Forstwirthschaftliche Zeitung*, ed. von Joseph Arnstein VII, Nr. 11, March 14 (1857), 180.

Anonymous (1858). Die allgemeine Versammlung der k. k. Landwirthschafts-Gesellschaft in Wien am 5. und 6. Mai 1858. *Allgemeine Land- und Forstwirthschaftliche Zeitung*, 8:19, 290–300, esp. 300 und Nr. 20, 305.

Anonymous (1866). Musikleben in Rußland. *Blätter für Musik*, 6:41.

Bachl-Hofmann, C., Cernajsek, T., Hofmann Th., and Schedl. A. (eds.) (1999). *Die Geologische Bundesanstalt in Wien. 150 Jahre Geologie im Dienste Österreichs (1849–1999)*. Vienna: Böhlau Verlag.

Baltzarek, F. (1980). Das territoriale bevölkerungsmäßige Wachstum der Großstadt Wien im 17., 18. und 19. Jahrhundert. Mit Betrachtungen zur Entwicklung der Wiener Vorstädte und Vororte. *Wiener Geschichtsblätter 35*, 1–30.

Bateman, F. (1855). On the Present State of our Knowledge on the Supply of Water to Towns. In: *Report of the Eighteenth British Association for the Advancement of Science*. London: John Murray, 62–78.

Benjamin, W. (1991). Das Passagen-Werk. In: Tiedemann, R. (ed.), *Gesammelte Schriften Walter Benjamins* (Vol. 5). Frankfort am Main: Suhrkamp.

Brantz, D., Disko, S., and Wagner-Kyora, G. (eds.) (2012). *Thick Space: Approaches to Metropolitanism*. Bielefeld: Transcript.

Coen, D. R. (2018). *Climate in Motion: Science, Empire, and the Problem of Scale*. Chicago: University of Chicago Press.

Darwin, Ch. (1859). *On the Origin of Species and the Preservation of Favoured Races by Natural Selection*. London: John Murray.

Feichtinger, J. (2010). *Wissenschaft als reflexives Projekt. Von Bolzano über Freud zu Kelsen: österreichische Wissenschaftsgeschichte 1848–1938*. Bielefeld: Transcript.

Feichtinger, J. (2012). 'Staatsnation', 'Kulturnation', 'Nationalstaat': The Role of National Politics in the Advancement of Science and Scholarship in Austria from 1848 to 1938. In: Ash, M., and Surman, J. (eds.), *The Nationalization of Scientific Knowledge in the Habsburg Empire, 1848–1918*. Basingstoke: Palgrave Macmillan, 57–82.

Gingrich, S., Haidvogl, G., and Krausmann, F. (2012). The Danube and Vienna: Urban Resource Use, Transport and Land Use 1800–1910. *Regional Environmental Change* 12:2, 283–294.

Glatter, E. (1864) *Beiträge zu einer ärztlichen Topographie Wiens mit besonderer Berücksichtigung der Mortalität im Jahr 1862*. Vienna: Leopold Sommer.

Greene, M. T. (1982). *Geology in the Nineteenth Century: Changing Views of a Changing World*. Ithaca and London: Cornell University Press.

Grimaud de Caux, G. I. (1839). *Considérations hygiéniques sur les eaux en general et sur les eaux de Vienne en particulier.* Paris: Imp. De Delanchy.

Grimaud de Caux, G. I. (1860). *Mémoire sur les eaux de Paris. Project de distribution Générale.* Paris: Racon.

Haidinger, W. (1850). Programm. *Jahrbuch der k. k. Geologischen Reichsanstalt* 1, V–VII.

Hauer, F. (1859). *Diary.* Manuscript in the Library and Archive of the Geologische Bundesanstalt, Vienna.

Hauer, F. (1882). Todes-Anzeige. Heinrich Wolf. *Verhandlungen der k. k. geologischen Reichsanstalt.* Bericht vom 31. Oktober 1882, Nr. 14, 253–255.

Hofbauer, G. (2003). Geognosie: Von Füchsels reflektierter Spekulation zu Werners voreingenommener Wahrnehmung. In: Albrecht, H., and Ladwig, R. (eds.), *Abraham Gottlob Werner and the foundation of the Geological Sciences.* Freiberg: Technische Universität Bergakademie, 164–176.

Hofmann, Th., Blöschl, G., Lammerhuber, L., Piller, W. E., and Şengör, A. M. Celăl (eds.), (2014). *The Face of the Earth: The Legacy of Eduard Suess.* Vienna: Edition Lammerhuber.

Hofmann, Th., and Klemun, M. (eds.) (2012). *Die k. k. Geologische Reichsanstalt in den ersten Jahrzehnten ihres Wirkens. Neue Zugänge und Forschungsfragen* (Berichte der Geologischen Bundesanstalt, 95). Vienna: Geologische Bundesanstalt.

Jacquin, J. F., and Partsch, P. (1831). *Die Artesischen Brunnen in und um Wien. Nebst Geognostischen Bemerkungen über dieselbe.* Vienna: Carl Gerold.

Karrer, F. (1877). Geologie der Kaiser Franz Josefs Hochquellen-Wasserleitung: eine Studie in den Tertiärbildungen am Westrande des Alpinen Theiles der Niederung von Wien. *Abhandlungen der k. k. Geologischen Reichsanstalt* 9:13, 1–420.

Karrer, F. (1886). Die Monumetalbauten in Wien und ihre Baumaterialien. *Ausserordentliche Beilage zu den Monatsblättern des Wissenschaftlichen Club in Wien* 7, 51–61.

Kennedy, C., Cuddihy, J., and Engel-Yan, J. (2007). The changing metabolism of cities. *Journal of Industrial Ecology* 11:2, 43–59.

K. K. Statistical Central Commission (1864). *Statistisches Jahrbuch der Oesterreichischen Monarchie.* Vienna: Kaiserl.-königl. Hof- und Staatsdruckerei in Commission bei Pfandel & Ewald.

Klemun, M. (2009). "Da bekommen wir auf einmal wieder zwei Etagen mehr! Wohin soll das noch führen!" Geologische Wissenskommunikation zwischen Wien und Zürich: Arnold Escher von der Linths Einfluss auf Eduard Suess' alpines Deckenkonzept, diskutiert anhand seiner Ego-Dokumente (1854–1856) und seiner Autobiografie. In: Seidl, J. (ed.), *Eduard Suess.* Vienna: VR unipress, 295–318.

Klemun, M. (2011). Understanding of Resources and Knowledge of Raw Materials, as Presented at the Big Exhibitions in the 19th Century. In: Ortiz, J. E., Puche, O., Rábano, I., and Mazadiego, L. F. (eds.), *History of Research in Mineral Resources.* Madrid: Cuadernos del Museo Geominero 13, 247–252.

Klemun, M. (2012). National, Consensus' as Culture and Practice: The Geological Survey in Vienna and the Habsburg Empire (1849–1867). In: Ash, M., & Surman, J. (eds.), *The Nationalization of Scientific Knowledge in the Habsburg Empire, 1848–1918.* Basingstoke: Palgrave Macmillan, 83–101.

Klemun, M. (2015). Geognosie versus Geologie: Nationale Denkstile und kulturelle Praktiken bezüglich Raum und Zeit im Widerstreit. *Berichte zur Wissenschaftsgeschichte. Organ der Gesellschaft für Wissenschaftsgeschichte* 38:3, 227–242.

Klemun, M. (2017). Spaces and Places: An Historical Overview of the Development of Geology in Austria (Habsburg Monarchy) in the Eighteenth and Nineteenth Centuries. In: Mayer, M., Clary, R. M., Azuela, L. F., Mota, T. S., and Wołkowicz (eds.), *History of Geoscience: Celebrating 50 Years of INHIGEO*. London: Geological Society, Special Publications, 442, 263–270.

Klemun, M. (2018). Different Functions of Learning and Knowledge—Geology Takes Form: Museums in the Habsburg Empire, 1815–1848. In: Rosenberg, G. D., and Clary, R. M. (eds.), *Museums at the Forefront of the History and Philosophy of Geology: History Made, History in the Making*. Boulder, CO: The Geological Society of America, Special Paper 535, 163–175, https://doi.org/10.1130/2018.2535(10).

Klemun, M. (2020). *Wissenschaft als Kommunikation in der Metropole Wien. Die Tagebücher Franz von Hauers der Jahre 1860–1868*. Vienna, Cologne, and Weimar: Böhlau Verlag.

Knell, S. (2000). *The Culture of English Geology, 1815–1851: A Science Revealed through Its Collecting*. Aldershot: Ashgate.

Kretschmer, W. (1999). *Geschichte der Weltausstellungen*. Frankfurt and New York: Campus Verlag.

Malthus, T. R. (1826). *An Essay on the Principle of Population*. London: John Murray.

Mayer, W. (1972). Gebietsänderungen im Raume Wien 1850–1910 und die Debatten um die Entstehung eines Generalregulierungsplanes von Wien. Phil. dissertation, University of Vienna.

Ragsky, F. (1848). Chemische Analyse des Wassers aus dem artesischen Brunnen des Herrn Rüdelmann bei der Mariahilfer-Linie. *Haidingers Berichte* 3, 90–92.

Reif, H. (2012). Metropolises: History, Concepts, Methodologies. In: Brantz, D., Disko, S., and Wagner-Kyora, G. (eds.), *Thick Space: Approaches to Metropolitanism*. Bielefeld: Transcript, 31–48.

Roskamm, N. (2011). *Dichte: Eine transdisziplinäre Dekonstruktion. Diskurse zu Stadt und Raum*. Bielefeld: Transcript.

Rudwick, M. (1963). The Foundation of the Geological Society of London: Its Scheme for Co-Operative Research and Its Struggle for Independence. *The British Journal for the History of Science* 4, 325–355.

Rudwick, M. (1996). Minerals, Strata and Fossils. In: Secord, J. A., and Spary, E. C. (eds.), *Cultures of Natural History*. Cambridge: Cambridge University Press, 266–286.

Schübl, E. (2010). *Mineralogie, Petrographie, Geologie und Paläontologie. Zur Institutionalisierung der Erdwissenschaften an österreichischen Universitäten, vornehmlich in jener in Wien, 1848–1938* (Scripta geo-historica 3). Graz: Grazer Universitätsverlag Leykam.

Simmel, G. (1903). Die Grossstädte und das Geistesleben. In: Petermann, Th. (ed.), *Die Grossstadt. Vorträge und Aufsätze zur Städteausstellung* (Jahrbuch der Gehe-Stiftung 9). Dresden: Jahn und Jaensch, 185–206.

Smith, R. A. (1849). On the Air and Water of Towns. In: *Report of the Eighteenth British Association for the Advancement of Science*. London: John Murray, 16–31.

Stur, D. (1869). Die Bodenbeschaffenheit der Gegenden südöstlich bei Wien: ein Bericht über die, der Gemeinde Wien zur Anlage eines Centralfriedhofes offerirten Flächen in den Gemeinden: Kaiser-Ebersdorf, Rannersdorf, Himberg, Pellendorf und Gutenhof, in: *Jahrbuch der k. k. Geologischen Reichsanstalt* 19, 465–484.

Sueß [*sic!*], E. (1916). *Erinnerungen.* Leipzig: Hirzel.

Suess, E. (1862a). *Der Boden der Stadt Wien nach seiner Bildungsweise, Beschaffenheit und seinen Beziehungen zum bürgerlichen Leben: Eine geologische Studie.* Vienna: Braumüller.

Suess, E. (1862b). Bemerkungen über die Einführung des geologischen Unterrichts an unseren Gymnasien. *Zeitschrift der österreichischen Gymnasien* 13, 165–177.

Suess, E. (1875). *Die Entstehung der Alpen.* Vienna: Wilhelm Braumüller.

Suess, E. (1888). *Das Antlitz der Erde* (2 Vols.). Prague, Vienna, and Leipzig: F. Tempsky.

Suess, E. (^{2}1892). *Das Antlitz der Erde* (2 Vols.). Prague, Vienna, and Leipzig: F. Tempsky.

Suess, E. (^{3}1909). *Das Antlitz der Erde* (2 Vols.). Prague, Vienna, and Leipzig: F. Tempsky.

Tollmann, A. (1963). Hundert Jahre Geologisches Institut der Universität Wien (1862–1962). *Mitteilungen der Gesellschaft der Geologie- und Bergbaustudenten in Wien* 13, 9–40.

Wagner-Rieger, R. (ed.) (1972–1981). *Die Wiener Ringstraße. Bild einer Epoche* (Vols. I-XI). Wiesbaden: Steiner Verlag.

Weigl, A. (2000). *Demographischer Wandel und Modernisierung in Wien (Kommentare zum Historischen Atlas von Wien 1).* Wien: Pichler.

Wolf, H. (1870). Neue geologische Aufschlüsse in der Umgebung von Wien durch die gegenwärtigen Eisenbahnarbeiten. *Verhandlungen der k. k. Geologischen Reichsanstalt*, 139–147.

Wolf, H. (1874). Die Gesteine des Gotthard-Tunnels. *Verhandlungen der k. k. Geologischen Reichsanstalt*, 140–145.

Wolman, A. (1965). The Metabolism of Cities. *Scientific American* 213, 179–190.

Part Three

Comparative Studies and Metropolitan Networks, 1870s–1910s

6 Polar Waters in Metropolitan Space

Circulating Knowledge about the Ice-Free Arctic Ocean in Hamburg and Vienna

Ulrike Spring

Introduction: Spaces of knowledge

Whilst images of melting polar ice have become part of everyday media reporting today, in the late nineteenth century the question was whether an open Arctic Ocean existed at all. Was there navigable water beyond the ice barrier? And if so, would this pave the way to one of the great enigmas of exploration history, the North Pole? Questions like these were vividly discussed in the media, in academic circles, and in geographical societies, and they received regular input through the reports of the many polar expeditions of the time. As they do today, expert mingled with popular knowledge, eyewitness reports with armchair analyses. In this chapter, I examine some of these knowledge forms as they emerged in the wake of the two Austro-Hungarian Arctic expeditions in the early 1870s, with a particular focus on the metropolitan space of Vienna.

Researchers in the history of science and knowledge have in recent years increasingly become interested in the question of how scientific knowledge is communicated within and across different spaces such as lecture halls or press media (Daum 2002; Lightman 2007; Livingstone and Withers 2011). Most works can be connected to Henri Lefebvre's observation that spaces produce social relations and are themselves a product of these relationships; they do not just exist by themselves and instead are shaped by their historical and societal contexts, actors, objects, and institutions, political, cultural, and social relations (Lefebvre 1992; see also Ash 2000, 2006; Sarasin 2011). Knowledge moves between these spaces and is transformed in the process.

This approach is highly useful in understanding how spaces in which knowledge is communicated interconnect or even intersect and how these spaces and the knowledge mediated there are shaped by a multitude of external and internal factors. However, a disadvantage of the approach is that it can rely too much on the idea of spaces as definable variables. The recently formed 'blue humanities' research perspective, following on the more solidly established 'green humanities', can account more strongly for fleeting relations *across* spaces, and is particularly important for understanding late nineteenth-century discussions about the open Arctic Ocean. Within 'blue humanities', the ocean is conceptualized not only as a specific geographical space, but more importantly as an integral part of human knowledge in general (see Buchanan 2018). In the context of this chapter, this means that one needs to look at the Arctic as a global space that has had an impact far beyond its geographical location and permeated different spaces of knowledge production and communication in different ways. In the nineteenth century polar exploration was a central part of Western global colonialism and imperialism. We also have to remember that the concept of 'the Arctic' is a cultural construct with flexible borders, its definition depending on interpretation (Keskitalo 2002; Holtorf 2017, 222).

Hence, while my focus is on spaces of polar knowledge communication and the circulation of such knowledge in Vienna—and to some extent Hamburg—my starting point is that the Arctic was an integral part of these urban spaces in the 1870s (for Vienna examples see also the chapters by Klemun and Mattes, Chapters 5 and 9 in this volume), although its role changed according to context. Howes (2013) shows the difficulties traveler-naturalists encountered when translating fieldwork experiences and vocabulary into metropolitan discourses. As the following examples illustrate, polar explorers met similar challenges on their return, with different spaces providing more authority than others, and different audiences allowing for more or less use of evidence.

The concept 'circulation of knowledge' has met with growing interest within the humanities and the social sciences in recent years. Whilst there is no consensus on what the concept precisely entails, its potential lies particularly in its dynamic open-endedness and potential for continuous alteration (see Östling et al. 2018 for an overview). I interpret the term not only as designating the movement and changing of knowledge from one place to another as time passes, but also as including detours, interruptions, changes, and anachronisms. Crucially, knowledge is continuously produced anew and interacts with the materiality of its surroundings, with the expectations of readers and discussants. It comes about within a highly complex web of configurations, which do not necessarily follow external logics. The term *entanglement*, inspired by the *histoire croisée* (Werner and Zimmermann 2006), may therefore be a useful supplement to that of *circulation*. In order to understand this entanglement of knowledges, I will investigate some physical spaces such as lecture halls and press media spaces in which knowledge of the Arctic briefly intensified in the early 1870s, when the two Austro-Hungarian polar expeditions took place. As Withers and Livingstone (2011, 4) point out, the nineteenth century is particularly interesting for analyzing such spaces, since science then became a 'public good'. Science was performed through new and heterogeneous audiences and sites and circulated via a multitude of different strategies.

The Austro-Hungarian polar expeditions under the leadership of officers Carl Weyprecht and Julius Payer are fascinating case studies, because they transferred the Arctic both metaphorically and literally to Central Europe and contributed briefly to an increased interest in and knowledge of all things Arctic among a scientific and a non-scientific audience. I am particularly interested in understanding how this knowledge was circulated through a highly complex and entangled performance of bodies, names, genres, images, objects, narratives, and settings in specific places. The urban spaces of the city, but also its premises—the ability to print, to employ journalists, to house many people in lecture halls, etc.—are central preconditions for understanding how such performances evolved.

1. The Arctic in Europe

When the second Austro-Hungarian polar expedition returned to Europe in September 1874 after more than two years in the Arctic, the discussion of the nature of the North rapidly intensified in popular and semi-academic media and forums all across Europe, but particularly in Vienna, as the city which had been most active in giving financial and moral support to the expedition and where the expedition officially ended and was dissolved. In the metropolitan space of Vienna, one of the largest cities in Europe, imperial residence and capital of the Austrian part of the Habsburg Empire, knowledge was produced and communicated across multiple media and built upon a wide array of Arctic images (for a detailed analysis of the reception of the expedition, see Schimanski and Spring 2015). The Arctic had become part of popular culture in Europe and North America since the early 1800s as a result of numerous and often tragic British and North American polar expeditions, of exhibitions on Arctic topics including its indigenous population, and of panoramas and dioramas at fairs (Potter 2007). In addition, Viennese public knowledge in 1874 could build upon experience acquired closer to home. Julius Payer had participated in the Second German North Polar Expedition (1869–70), and both he and Weyprecht had conducted a pre-investigative expedition in the summer of 1871 into the sea around Spitsbergen. In the spring of 1872, a public exhibition on the planned main expedition took place in Vienna. In 1874, after the expedition's return, the two leaders gave short talks and held lectures in the Norwegian city of Bergen and the Geographical Societies in Hamburg and Vienna, and the Viennese newspapers published thousands of articles on the story of the expedition and the enormous celebrations of its participants across Europe. As a result, Northern space was folded into the metropolitan Central European space of Vienna.

Different spaces of knowledge communication entailed different physical materialities and connotations, assigning them different degrees and forms of evidence and authority, and in turn influencing the way scientific knowledge was communicated there. A space conveyed authority through its societal status: A lecture in the Academy of Science provided a different setting than a speech at a banquet, and the serious and authoritative newspaper *Neue Freie Presse* carried a different weight than the sensationalist *Neuigkeits-Welt-Blatt*. Status was also signaled by spaces' architecture and design, their accessibility, but also by the people using them, along with the instruments of which they availed themselves, including the genre the journalist wrote in or the objects the lecturer displayed in order to illustrate his talk, such as maps, natural or other artifacts. The urban context as a space in which science could be performed publicly to a wide range of people played an important role in this process—in larger metropolitan cities a more varied knowledge accumulation and production was in place than in smaller cities—but also the geographical distance to the North has to be taken into account. The reception celebrations in the metropolis Vienna displayed the North and the collected knowledge there in a far more spectacular and speculative manner and across many more media than did the more scientifically focused reception in the much smaller Norwegian city of Bergen, with its long polar hunting tradition, or the enthusiastic yet ordered welcome in the German maritime city Hamburg (Schimanski and Spring 2009).

Such a focus on geographical relationality complicates the chronology of events. While one way to approach the communication of polar knowledge in various media and geographical spaces would be to follow the expeditions' footsteps from their arrival in Norway all the way to Vienna and to trace the increase of knowledge during this journey, this would not explain various detours and delays. On arriving in Vardø in September 1874, the exhibition leaders sent telegrams to Vienna, Hamburg, and Fiume (today's Rijeka) among other places, initiating a debate about a potentially open Arctic Ocean in these urban spaces long before the expedition members physically arrived there. When the explorers reached Hamburg, journalists and scientists had already had time to elaborate their views in various newspapers. Whereas we as researchers today (partly thanks to digital technology) have an overview over these discussions and their sequence, very likely no one was in that position in the 1870s. This is another reason why I argue for the term *entanglement* to be added to that of *circulation*.

The spaces I discuss here are only a few of many spaces of public reception in Vienna that developed in the 1870s. Other important spaces were the overwhelming mass spectacle in Vienna on the second expedition's arrival at the Nordbahnhof railway station, the theatres and banquets with their many laudatory speeches, and the rich material of scientific articles and books published in the aftermath of the expeditions (Schimanski and Spring 2015).

2. An Ice-Free Arctic Ocean

The question of the nature of the North and in particular the geographical North Pole, particularly whether there was open water or land at the cap, had for long been the subject of lively discussion in the scientific and explorer communities in Europe and Northern America (Bravo 2019; Luedtke 2015; Holtorf 2014). Holtorf (2012, 190–192) discerns a tension in the mid-nineteenth century over this question between popular speakers, amateur scientists, and enthusiastic polar explorers on one side, and an increasingly skeptical yet curious scientific community on the other. After the expeditions returned, the discussion continued in the salons, the streets, and in Vienna's many taverns. In the later part of the nineteenth century the discussion was embedded in several interrelated discourses, the most dominant being those of science, imperialism, and economics—exploitation of natural resources but also ship traffic. Scientists wished to learn about the processes behind climatic and atmospheric riddles on a global basis. Weyprecht in particular saw expeditions to the North as an opportunity to find out more about such issues and later initiated what came to be called the First International Polar Year (1882–1883), an international scientific research collaboration taking place in the Arctic and the Antarctic. At the same time science played a central role in nineteenth-century nation-building processes, as scientific knowledge was often equated with a culture's state of civilization (for a discussion of this process in the Habsburg Empire see Ash and Surman 2012). Closely related were the imperial and colonial discourses, referring to Western countries' desire to map the globe. Widespread interest in German and Austrian polar research started in the 1860s and has to be seen as part of these discourses (Walsh 2016; for general overviews see Krause 2010; Murphy 2002). While toward the end of the nineteenth century the Pole would become an aim in itself (Bravo 2019), in the 1870s the distinction between the Pole and the still little explored lower regions was unclear. In popular imagination and in the discussion of open waters in the Arctic these geographical spaces often converged. The public debate about the open sea thus was in continuous interplay with prevalent knowledge and conceptions of the North Pole.

When Weyprecht and Payer returned to Tromsø in Norway with their companions after their summer-long investigative expedition in early October 1871, they sent a telegram to Vienna about having sighted open waters east of Spitsbergen. In the telegram they suggested that open water might extend to the well-known polynia (open water within the pack ice) close to the New Siberian islands in the east, and that this might be the best way of reaching the North Pole (Oesterreichische Nordpolfahrer 1871, 55). For the prominent German geographer and cartographer August Petermann, these findings confirmed his belief in a Gulf Stream extending far beyond the coast of Norway, and the existence of ice-free waters north of 79° latitude. Petermann was the central initiator of the German and Austrian polar activities from the 1860s onwards (Abel and Jessen 1954), and a well-known propagator of the open Arctic Ocean theory. The path to the North chosen by the two Austro-Hungarian expeditions was based on Petermann's thesis that the best way of going North was via the European Arctic, between Spitsbergen and Novaja Zemlja, in contrast to most American and British polar expeditions, which had chosen the American Arctic as their starting point (Felsch 2010, 105–108). Petermann showed his satisfaction in an article in the official *Wiener Zeitung* already a few days after the expedition's return, and reminded his readers of a map he had published in June 1870, where the name "Gulf Stream" and an arrow pointed exactly to the point which "the equally competent, scientific and prudent sea officer Weyprecht" had written about in his telegram (Petermann 1871a, 180; this and the following translations by the author). Direct observation in the field as well as character traits were used to bolster his claim, providing evidence through the authority of the media used: body and map.

Petermann's satisfaction also has to be seen in the context of a heated debate about his scientific contributions (and implicitly, his personality) preceding the two Austro-Hungarian expeditions, involving the leader of the Second German Polar Expedition Carl Koldewey in particular. Koldewey had earlier expressed his disbelief that any ship could cross the ice between Spitsbergen and Novaja Zemlja toward the North Pole, adding that he would only embark on such a journey if Petermann himself would join him (Koldewey 1871, 92)—an implicit criticism of Petermann for never having been to the Arctic himself. Petermann was indeed a so-called armchair scientist, his theories depending on knowledge acquired by others in the field, and an effort to acquire funding for his own expedition in 1872 proved futile (Felsch 2010, 219). In October 1871, however, his theory seemed to have been proven true, and he used Payer and Weyprecht's telegram and the scientific character of their expedition to dismiss Koldewey's statement in the letter to the *Wiener Zeitung* mentioned previously (Petermann 1871a, 179). By calling on Weyprecht and Payer, two men who actually had been to the North and seen for themselves, Petermann could merge field observation and scientific knowledge to produce the ultimate authoritative statement (for the debate see Felsch 2010, 208–210, 216–217).

Petermann's map and comments were published in the leading German geographical journal of the time—edited by himself and popularly called *Petermann's Geographische Mittheilungen* (e.g. Petermann 1871b)—but also in leading newspapers, thus opening the debate on an ice-free ocean to a wider audience, and inserting it into the public spaces of Vienna. Payer would soon confirm his support of Petermann's thesis in a short letter addressed to the Association for Geography and Statistics in Frankfurt am Main in 1871 (Payer 1871), and published in the *Wiener Abendpost*, the evening issue of the official *Wiener Zeitung*, on October 27. In it, he also used sightings by Norwegian skippers as proof for the existence of open waters in summer, and argued for the disappearance of "an enormous territory of ice from our maps" (Von der Polar-Expedition 1871, 992). He would emphasize his and Weyprecht's sightings of an open sea in a talk held in the Geographische Gesellschaft in Vienna in late November 1871, and subsequently the Society passed a resolution supporting the plan of an extended expedition (Oberlieutenant Payer 1871, 4–5). Similarly, the Academy of Science was to use the belief in open waters as a main argument for supporting the 1872 expedition in an assessment they had conducted on its possible benefits (Expedition . . . 1872). The confidence of Payer in the new discovery has very likely also to be seen in the context of his efforts to acquire funding for the new expedition. The financial support for the expedition was overwhelming indeed, with the polar explorers managing to mobilize contributors from all social classes, many of them from Vienna (Rechnungsabschluß 1874, 1104–1106). Also in this respect, Vienna became the center of the expedition's ambitions.

Throughout the months between the return of the 1871 expedition and the departure of the 1872 expedition, reports in Viennese newspapers suggest a growing confidence in the existence of an ice-free Arctic Ocean. Moreover, with the discovery of open waters and easy navigability through the ice fields north of Novaja Semlja, Austria-Hungary had contributed to solving one of the great enigmas of the world. Already in June 1871, before embarking on the first expedition, Weyprecht had asked Petermann—if it were at all necessary to make such statements—to propagate the expedition as Austrian, not German, as it was the Austrians who had funded the 1871 expedition and who were expected to provide them with enough funding for a new expedition (Berger et al. 2008, 333). Although Weyprecht in his very next sentence pointed out that he and Payer considered science to be international (ibid.), his wish illustrates that he was aware of how closely interlinked science and national status were at the time. Accordingly, one of the Austrian Academy of Science's reasons for recommending the 1872 expedition was that it might help "to secure Austria the glory of important geographical discoveries" (Expedition . . . 1872). The findings thus had a direct impact on Vienna, not in a concrete economic way, as would have been the case in a city closer to the Arctic such as Bergen or even Hamburg, but in a symbolic way. Vienna became the signifier of scientific Austria: The Geographical Society referred in its public call for financial support for the expedition to "rich Vienna that upholds science" and which managed to cross the gap between mind, science, and materialism (Aufruf . . . 1872, 6). The Society here addressed another topic that was prominent throughout the reception of the expeditions and intensified around the expeditions: The focus was on idealism and altruism, both associated with science. This argument was a response to the question why a Southern state such as the Habsburg Empire should support a journey to the North, which was dangerous, expensive, and of little concrete use for the country. The expeditions were seen as a proof that idealism existed in the world and that materialism had not completely succeeded (see Spring and Schimanski 2015). The question of an ice-free Arctic Ocean was essential in this respect, as it referred to knowledge that was not of immediate relevance to the Viennese or Austrians more generally; it hence could affirm the monarchy's disinterested contribution to global science. Vienna represented this role both on a symbolic and an actual level through its status as a Central European metropolis and imperial capital.

The open sea had yet another symbolic dimension which we need to keep in mind. As Johan Schimanski and I have shown earlier (2019), word plays with ice flourished after the expedition's return in 1874, juxtaposing the experience of the Viennese with ice (after all, Viennese knew well the dangers of ice, flooding, and ice jams on the Danube) with that of the polar explorers in the Arctic. The fact that the expedition members had conquered the slippery and unreliable world of the ice, had besieged the mythical place of the North, and even contributed to understanding the balance between an ice-free and ice-covered ocean, proved their superior standing and assertiveness.

There were also some concrete material findings which found their way to Vienna from open polar waters and which literally brought the Arctic into the city: The expedition members brought back driftwood from their journey in 1871, and the Academy of Sciences, under the leadership of Professor Julius Wiesner, could establish that it was wood from trees from North Asia. According to the newspaper article, it was the first time that wood from such a high Northern region had been investigated (Treibhölzer . . . 1872, 4). In 1874, the expedition members also brought various animal and plant samples back to Vienna, many of which are still to be seen in the Natural History Museum there.

3.   Disclaiming the Existence of an Open Arctic Ocean

However, as it turned out, 1871 had had better ice conditions than 1872, again changing knowledge about the North and showcasing the variable and unreliable character of ice more generally and of the Arctic in particular. When the expedition left Tromsø on board their ship the *Tegetthoff* in June 1872, the plan was to explore the sea in the northeast of Siberia, traveling via the northern edge of Novaja Zemlja. However, shortly after leaving Norway, they were caught in the ice, drifting to the northwest instead of to the east, and eventually having to abandon their ship. The reception on their return in September 1874 was much more triumphant than in 1871, mainly because of the discovery of a group of islands northeast of Spitsbergen, which they called Franz Josef Land. The space of Vienna was quickly filled with self-named experts, glacial metaphors, and

Northern souvenirs. One could buy Arctic clothes and Arctic sweets in the streets of Vienna, Austrian food acquired Arctic names in the welcoming banquets, and journals published caricatures featuring the North (Schimanski and Spring 2015, 420–427). As a result, the question of an ice-free Arctic Ocean became entangled with other parts of the Arctic discourse emerging in Vienna at the time. At the same time, Petermann's thesis of an open ocean was this time seriously challenged. Actual observation would now counter and not confirm his ideas. A gap between Petermann's representation of the field and the experience of being in the field emerged, leading to conflicting interpretations of the nature of the Arctic.

4. Mapping Arctic Ice

A few months before embarking on their journey in 1872, the leaders of the expedition organized an exhibition, mainly displaying information about their plans and their equipment (Die Ausstellung der Nordpol-Expedition 1872, 4). The venue was the Kursalon in Vienna, a site often used for festive upper-class events and which hosted the celebratory banquet for the officers when they returned. One of the illustrations on the exhibition (Figure 6.1), which was reprinted in September 1874 in a brochure (Singer 1874, 39), hence becoming part of the 1874 discourse as well, showed a conglomeration of polar objects.

The map shows the wide area covered by the term 'North Pole', with the Pole being the center of the map and the map being the center of the image. The white of the North is not distinguished from the white of the sea, leaving it open to the viewer's imagination whether there is ice or water covering the Pole region. Even though the newspapers in their reviews of the exhibition generally did not refer to the question of the open polar sea, we may assume that it functioned as a background of the reception of the exhibition. Holtorf (2014, 167) shows how maps of the Arctic were produced for different aims, but obtained an "intrinsic discursive value" for the onlooker. Maps reaffirmed or disturbed theories of the open polar sea, depending on their interpretation. Hence, this map and other tools of knowledge production were part of the debate over the nature of the North.

Before the expedition began, the Arctic, with its many unknown regions and the mysterious North Pole, was a space of projection for expectations and hopes as to what the Austrians would be able to achieve up there. The objects in the illustration were meant to inscribe Austrian competence into the white spaces and help conquer the wilderness of the North. When the expedition returned, the horizon of expectation would literally turn into a space of experience, to use Reinhart Koselleck's (1979) terms. While in 1872 the whiteness of the map represented a faraway horizon, with nobody knowing what was to lie beyond, to be filled only with anticipation, in 1874 part of it had Austro-Hungarian knowledge scribbled onto it. As the map in Figure 6.2 shows, the newly discovered Franz Josef Land

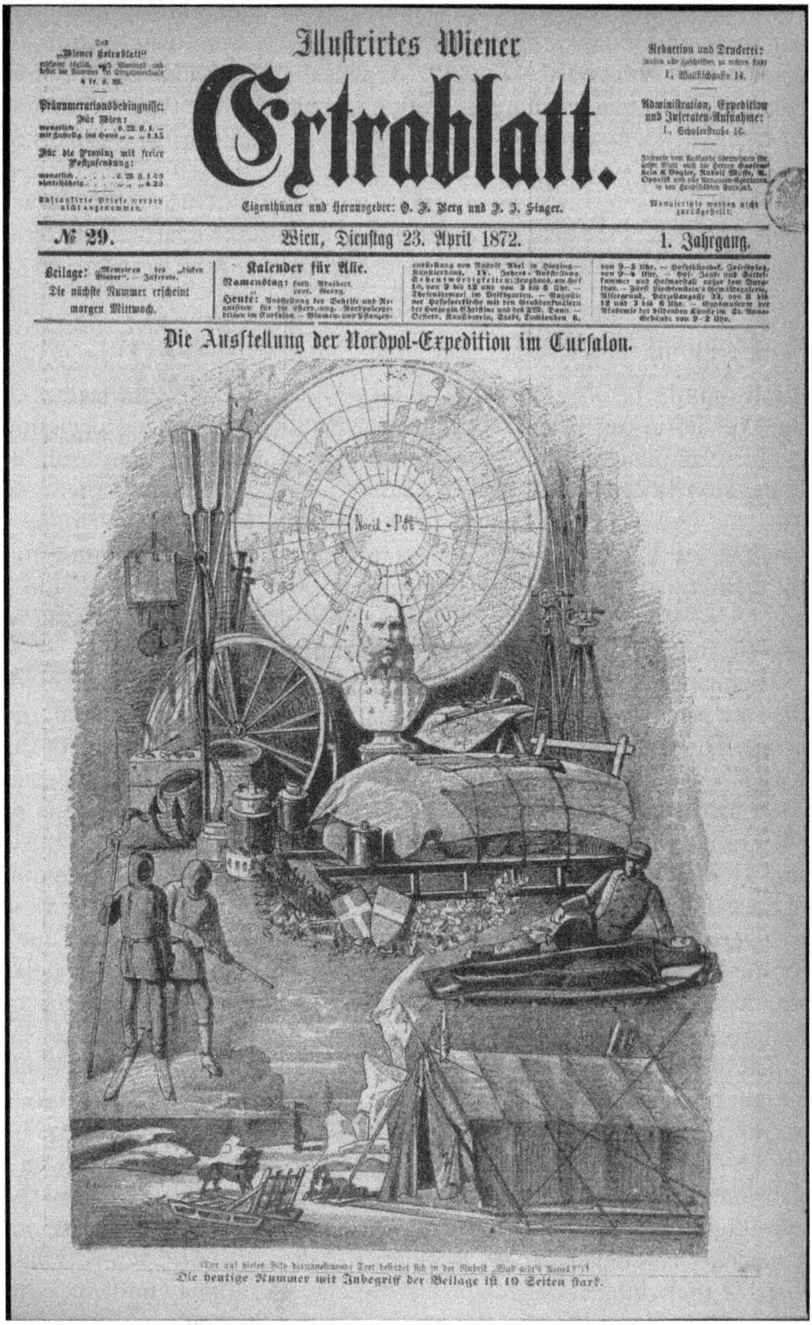

Illustrirtes Wiener
Extrablatt.

Eigenthümer und Herausgeber: O. F. Berg und F. J. Singer.

Redaction und Druckerei: I., Wallfischgasse 14.

Administration, Expedition und Inseraten-Aufnahme: I., Schulerstraße 16.

№ 29. Wien, Dienstag 23. April 1872. 1. Jahrgang.

Die nächste Nummer erscheint morgen Mittwoch.

Kalender für Alle.

Die Ausstellung der Nordpol-Expedition im Cursalon.

Nord-Pol

Die heutige Nummer mit Inbegriff der Beilage ist 10 Seiten stark.

Figure 6.1 The exhibition of the Arctic expedition in Vienna in the Spring of 1872, printed on the title page of the newspaper *Illustrirtes Wiener Extrablatt*. “Die Ausstellung der Nordpol-Expedition im Cursalon.” *Illustrirtes Wiener Extrablatt*, 1. Jahrgang, Nr. 29, April 23, 1872, 1.

Source: ANNO/Österreichische Nationalbibliothek.

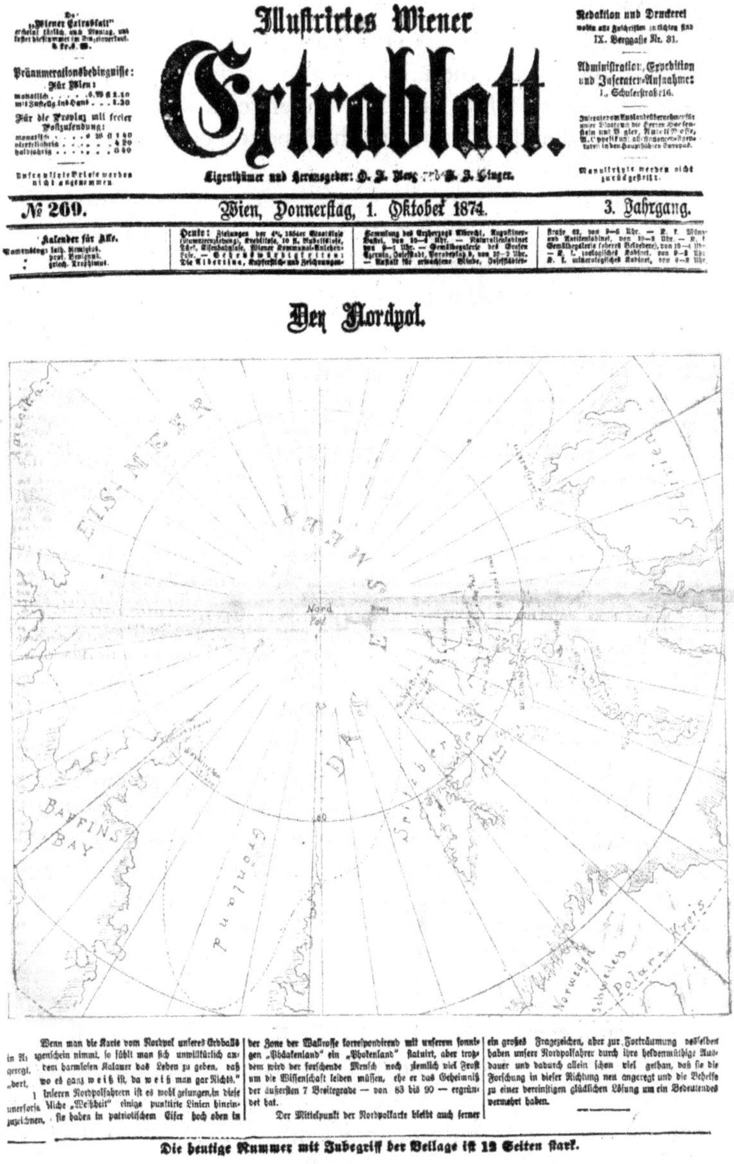

Illustrirtes Wiener
Extrablatt.

Redaktion und Druckerei
IX. Berggasse Nr. 31.

Administration, Expedition
und Inseraten-Aufnahme:
I., Schulerstraße 16.

Pränumerationsbedingnisse:
Für Wien:

Für die Provinz mit freier
Postzusendung:

Manuskripte werden nicht
zurückgestellt.

Eigenthümer und Herausgeber: O. F. Berg und F. J. Singer.

№ 269. Wien, Donnerstag, 1. Oktober 1874. 3. Jahrgang.

Kalender für Alle.

Der Nordpol.

Wenn man die Karte vom Nordpol unseres Erdballs in Augenschein nimmt, so fühlt man sich unwillkürlich angeregt, dem harmlosen Kalauer das Leben zu geben, daß „wo es ganz w e i ß ist, da w e i ß man gar Nichts." Unseren Nordpolfahrern ist es wohl gelungen, in diese unerforschliche „Weißheit" einige punktirte Linien hineinzuzeichnen, sie haben in patriotischem Eifer hoch oben in der Zone der Wallrosse correspondirend mit unserem sonnigen „Phäakenland" ein „Phokenland" statuirt, aber trotzdem wird der forschende Mensch noch ziemlich viel Frost um die Wissenschaft leiden müssen, ehe er das Geheimniß der äußersten 7 Breitegrade — von 83 bis 90 — ergründet hat.

Der Mittelpunkt der Nordpolkarte bleibt auch ferner ein großes Fragezeichen, aber zur Forträumung desselben haben unsere Nordpolfahrer durch ihre heldenmüthige Ausdauer und dadurch allein schon viel gethan, daß sie die Forschung in dieser Richtung neu angeregt und die Behelfe zu einer dereinstigen glücklichen Lösung um ein Bedeutendes vermehrt haben.

Die heutige Nummer mit Inbegriff der Beilage ist 12 Seiten stark.

Figure 6.2 On this map of Franz Josef Land published shortly after the expedition's return in *Illustrirtes Wiener Extrablatt*, the area around the North Pole is called "Eismeer". "Ice sea" refers not only to the German name of the Arctic Ocean, but also suggests the existence of ice rather than open water. "Der Nordpol". *Illustrirtes Wiener Extrablatt*, 3. Jahrgang, Nr. 269, October 1, 1874, 1.

Source: ANNO/Österreichische Nationalbibliothek.

reached close to the Pole. The North was in other words folded into the space of experience of the Viennese. The objects on display in the exhibition had in 1874 actually been to the North, literally displaying evidence. As these examples show, the discussion about the open waters was not just a scientific debate, but has to be understood within this wide field of Arctic knowledge that for a very short period entered urban space in autumn 1874.

The expedition's experience on Franz Josef Land had posed a serious challenge to the idea of an open Arctic Ocean. In his first extensive report from the expedition, a letter from the Norwegian steamer on their way back, which was published in the *Neue Freie Presse* on September 25, the day when the men returned to an enormous welcome in Vienna, Payer explicitly disclaimed any belief in an open Arctic Ocean, even stating that this had been so already before he departed in 1872. While he did observe open water off the coast of Franz Josef Land, he was of the opinion that this soon would be replaced by impenetrable pack ice. He confirmed his belief by loosely referring to "experiences and counter-evidence" that spoke against the thesis of an open Arctic Ocean (Payer 1874a, 1). Payer's use of terminology and the fact that this letter was the first longer eye-witness report by a member of the expedition underlined his argument and invested it with immediate authority. He carried with him corporeal evidence: his eyes had seen the sea and his body had been there. Here we can see a gap emerging between having been to and seen the North and merely representing it. This gap had also existed in 1871, but had temporarily been narrowed through similar views by the explorers and Petermann. While the letter was written in the private sphere on board a steamship off the Norwegian coast, it quickly became part of public space, being reprinted in local and international newspapers. In the following weeks, the two opposing positions—represented by Payer and Petermann respectively—were to shape the debate about the Arctic sea.

5.   The Question of Evidence

In the nineteenth century, bodily presence had much more relevance for conveying scientific evidence than it has today, in the age of live transmissions. Payer's next outspoken criticism of the open Arctic Ocean theory took place in the lecture hall of the Geographical Society of Hamburg on September 23, 1874 (for the reception in Bergen, see Schimanski and Spring 2015, 64–91, 430). While for us later readers, this was his second renunciation of the thesis after his letter from the steamer, for people living in 1874, this was the first time they had heard of it. Although Payer's talk was held in a scientific semi-private environment and only invited guests and members of the Society were allowed to attend, it almost immediately became part of the public sphere, because its text, or parts of it, was reprinted in newspapers across Germany and the Habsburg Empire, one day before the publication of the previously mentioned letter. Today, we have only access to this speech through this mediated form; the analysis of Payer's talk thus takes place from the perspective of the public space of the periodical press.

Various forms of scientific authority intersected here: Not only was Payer present in person, something which gave extra weight to his eyewitness account, but the sparsely decorated hall added a certain austerity to his remarks, well-tuned to scientific sobriety. According to the *Presse*, the room was "not very beautiful", only having Gothic windows, chairs, and a wooden ceiling as decoration (Die Nordpolfahrer 1874, 1). The lecture was also the first speech Payer gave about his sledge expeditions on Franz Josef Land. Taking Payer's rhetorical talent into account, this talk was in line with what Finnegan (2011, 155) identifies as an essential prerequisite in nineteenth-century oral dissemination of scientific knowledge: "science had to sound right as well as look right". While Finnegan makes this point in reference to intellectual culture in mid-nineteenth-century urban Britain, this no doubt also applied to public "speech spaces", to use his term (ibid.), such as that in Hamburg and later in Vienna. Moreover, the context of the Society provided legitimacy and gave weight to Payer's words, which his letter written in the isolation of a ship's cabin could not have done to the same extent.

According to the *Neue Freie Presse*, in the banquet preceding the Society's meeting, Payer gave a toast to the absent Petermann, calling him jokingly "his opponent" because of the latter's belief in the open Arctic Ocean (Die Nordpolfahrer in Hamburg 1874, 7). In the lecture on the following day, Payer acknowledged the existence of a non-frozen ocean in channels in the ice, although this did not lead him to believe that these could be used as waterways for ships. Neither did he observe any sign of drifting (Von der Nordpol-Expedition 1874, 8). When he and his men arrived at the northernmost point of their journey on Franz Josef Land, they could see lands extending in front of them even beyond 83° latitude; however, they also were convinced that "*this way* the North Pole was not *accessible* and the theory of an *open polar sea* was untenable" (ibid.; emphasis in the original). The *Presse* emphasized this argument by using different fonts, something that gives us an indication of the relevance of these new insights. While the *Presse* only reported Payer's talk, the *Neue Freie Presse* actively interceded in the controversy. We learn that in his response to Payer's talk, hydrographer and geophysicist Georg Neumayer of the Geographical Society in Berlin had pointed out that physicist and meteorologist Heinrich Wilhelm Dove, "the man who first had fought against the hypothesis of an open Arctic ocean, was present, and that his view had again been strengthened by Payer's examination" (Die Nordpolfahrer in Hamburg 1874, 8). Dove had published widely on questions of meteorology and was ascribed more scientific authority than the bustling Petermann (Felsch 2010, 219). The journalist of the *Neue Freie Presse* leaves the audience's or Payer's reaction unmentioned. However, in the extensive reporting on the expedition in the newspapers, we can find from time to time further suggestions of conflicts within the scientific community. Weyprecht's good friend Heinrich Littrow suggested in the *Presse* that the absence of Petermann in Hamburg could be explained by the presence of Koldewey, as "poles of the same name repel one another" (Littrow 1874, 1). Koldewey was then a resident of the city and an active member of the Geographical Society.

It was, however, more than a scientific disagreement that interested journalists, and presumably their readers, in autumn 1874. The contribution to the discussion of whether there was an open Arctic Ocean or not was for the time being, besides the discovery of Franz Josef Land, the only concrete outcome of the expedition, and it had also been one of its motivations. The perspective, however, had changed, from belief in Petermann's thesis in 1871 and 1872 to its dismissal in 1874. Petermann briefly turned from being an inspiration and supporter into an "opponent". The gap between eyewitness-evidence and armchair science widened and could not be closed at that point in time. Only a few days later, however, Payer's authority as an eyewitness would be called into question at the Extraordinary Meeting of the Geographical Society in the sumptuous Great Hall of the Academy of Sciences in Vienna on September 29, 1874 (see Schimanski and Spring 2015, 158–160). The main audience consisted of high-ranking members of the bourgeoisie, navy, army, and nobility, headed by the 16-year-old Crown Prince Rudolf, whose task it was to award Weyprecht and Payer with the Society's honorary diplomas (Die Nordpolfahrer in der Geographischen Gesellschaft 1874, 5). Here again, we can see the symbolic role Vienna was assigned as a representation of Austrian or even Austro-Hungarian achievements. The lectures held by the two expedition leaders would quickly be published in the Viennese newspapers and at the end of the year in *Petermann's geographische Mittheilungen* (Payer 1874b; Weyprecht 1874).

At the outset the stage was prepared to provide the speakers with scientific legitimacy. Payer showed two maps, of the North and Franz Josef Land respectively, thereby underlining the facticity of his talk. However, a member of the nobility allegedly expressed disbelief at Payer's report of his sledge expeditions, suggesting that the information about the newly discovered lands was untrue or at least exaggerated. At least as interesting as this denial of authority based on direct observation is the fact that we know about this jibe only by hearsay. It has almost become an urban legend in narratives about the expedition, being mentioned in most retellings of the expedition and in biographies of Payer; it is supposed to have been a direct contributing factor to Payer's decision to give up his career as an army officer and instead to turn toward art and become a prizewinning painter of Arctic scenes (Berger 2015, 136–137). Whereas this event hardly affected the overall picture of the expedition and its legitimacy, it shows that there existed skepticism in the Viennese population along-side the enthusiastic images conveyed in the newspapers (see also Spring 2018; Schimanski and Spring 2015, 174–178). We can only speculate whether this reported disbelief also was directed at or at least connected with Payer's outspoken criticism of the open Arctic Ocean thesis, but by becoming part of the overall reception of the expedition, it may also have affected the scientific and popular discussion about the ice-free ocean.

Payer's as well as Weyprecht's speeches in Vienna were to a certain extent repetitions of the ones in Hamburg, already known to the audience through newspaper reports. Whereas the venues of the two meetings provided similar backgrounds in terms of authority and legitimacy, the different contexts, geographical positions, and chronological order of the events, and most likely also the public way of dealing with the Arctic, affected their interpretation and reception. Certainly in Vienna, the scientific relevance of the speeches was toned down, with the journalist of the *Presse* stating that because of the illustrious audience present one could not expect the speeches to be reflections on the scientific results, but rather contributions to the festive atmosphere of the meeting (In der Geographischen Gesellschaft 1874, 7). Here, the symbolic role of Vienna as center of the Habsburg Empire becomes apparent. It could also well be that in Hamburg, where many had been to the North, the eyewitness had stronger scientific credibility than in Vienna, where the Arctic was a distant, spectacular, and mythical place, rather than a scientific and economically exploitable arena. In his talk, Payer again denied the existence of an open ocean, calling the open waters he had seen not more than a small polynia, and negated any possibility of reaching the North Pole via Franz Josef Land. He described discussion of these topics as being of peripheral interest and as "pointless speculation" (ibid.). The discourse had changed since 1871, when the topic had been considered of major significance.

6. Navigating Arctic Waters

There was continued discussion in the Viennese newspapers, mostly in the form of feuilleton columns. Journalist and geographer Joseph Chavanne and journalist, travel writer, and historian Friedrich von Hellwald were among the most active contributors. While both had cultural capital they could use to assert their positions, they represented different views, with Chavanne supporting Petermann and Hellwald trying to mediate between the opposite views (see also Schimanski and Spring 2015, 431). In his first feuilleton on the expedition in the *Neue Freie Presse* on September 19, Chavanne commended the explorers for having contributed to solving the riddle of how to reach the region around the North Pole: According to him, it had proved that the continuation of the Gulf Stream to Franz Josef Land and onwards to the Siberian polynia was an "undisputable fact". Referring to Petermann's and his own theories on the existence of ice-free waters, he praised the expedition as a rare example of theory and practice working in unison (Chavanne 1874a, 3). As in Petermann's letter from 1871, the difference between fieldwork and representation was dissolved and opened up for evidence based on reports rather than on eyewitnesses. Moreover, in Chavanne's closing words in the article in praise of German and Austrian achievements, he used science to reflect the enthusiastic mood in the Viennese streets, the belief in nationalized science, and Austria contributing to the progress of the world by helping to "unveil . . . the riddle of the Nordic sphinx" (ibid.).

In an article written on October 10 and published in *Petermann's Geographische*, Chavanne relied more heavily on Weyprecht as authority and stated that one aim of his text was to counter the many unscientific comments that had been put forward following the return of the expedition (Chavanne 1874b, 421). At the same time, however, he used the observations of the Austro-Hungarian expedition as well as those of other international expeditions to defend Petermann's thesis of a Gulf Stream continuing into the Northeast and of a temporarily ice-free sea (ibid., 421–425). In contrast to his article in the *Neue Freie Presse*, he did not appeal to national sentiment, and in general used more objectified language. Knowledge was cast in a different shape, meant to appeal to a different audience rather than responding to the Viennese reader and the huge public celebrations of the expedition across the city. His writings, however, led to an indignant response by Weyprecht, who sent a letter to the Geographical Society in Vienna on November 1, which was published in their *Mittheilungen* and referred to newspaper reports of the last meeting of the Society which indicated a consensus between his and Chavanne's perspectives on the ice-free ocean. Weyprecht's main argument was that this and other similar claims could have no basis in fact as long as he had not published his "views which were based on direct observation" (Berger et al. 2008, 422). He repeated this view in a letter to Petermann the same day, hoping that his perspectives would be published in the *Geographische Mittheilungen* (ibid., 422–423). In a later letter to Petermann, Weyprecht disclaimed Chavanne's authority in these questions, and even called him a "babbler" (ibid., 424).

Most journalists, however, readily accepted Payer's and Weyprecht's claims, suggesting that there was a dominating discourse of skepticism about the idea of the ice-free ocean already before the expedition's return. The *Illustrirtes Wiener Extrablatt* stated on September 5, only one day after news of the safe return of the expedition had arrived and when only brief fragments of the course of the expedition were available, that the "prevalent prejudgment of an open Arctic ocean and of all associated combinations have thoroughly and forever been disturbed by the discovery of Franz Josef Land" (Franz Josefs-Land! 1874, 2). While the newspaper, with its sensationalistic reportage and xylographs, presumably did not carry much weight in scientific circles (though it probably had much influence on the general public's perception of the Arctic region), the widely read scientific Hamburg journal *Hansa* did carry such weight, and advocated the same message in an article from October 18 (Zukünftige arktische Reisen 1874, 161).

In a third perspective, Friedrich von Hellwald sought to mediate between these two opposite opinions, also using Petermann's vague definition of the term "open polar sea" to support his argument. In a lengthy feuilleton from October 19 in the *Wiener Abendpost*, he argued that the debate had been marked by an inconsistent use of language, mixing up the idea of a *navigable* with that of an *open* Arctic Ocean. This was in fact a crucial observation for understanding the debate on an ice-free ocean, not only in the 1870s but from the eighteenth into the twentieth century (Luedtke 2015; Holtorf 2014). The Austro-Hungarian polar expedition had set out with the belief in a navigable, though not necessarily open Arctic Ocean, and, Hellwald implied, it was therefore no surprise that Payer came back as a "total opponent" to the idea of an open sea. He also cautiously criticized Payer for being too adamant in his views, specifically for stating that the path to the North Pole was closed from this side of the Arctic; for Hellwald, no scientific knowledge was set in stone, as long as there still were unknown questions and undiscovered places. For him, while there might not be a warm open ocean, there were many observations suggesting a partly navigable sea (Hellwald 1874, 1908). In this he followed Weyprecht, who had stated in a letter to Petermann already before their departure in June 1871 that he surmised that he would find navigable water behind a barrier of ice (Berger et al. 2008, 330). Hellwald kept to an academic approach, weighing pros and cons, while also speaking from the perspective of the theorist who, like Chavanne, tried to merge theory and practice. Payer's eyewitness status and the authority he acquired by bringing this knowledge into the urban spaces of Hamburg and Vienna were reduced here to being one report among many, which had to be scientifically analyzed and inserted into previous knowledge in order to gain validity.

In his hugely popular expedition report, published first in 1875 in installments and in 1876 as a book, Payer followed Hellwald's interpretation and defended Petermann: Petermann had "rather unjustifiably" been seen as the main advocate of the theory of an ice-free Arctic Ocean, when in fact all Petermann had argued for was the existence of an "inner Arctic ocean that would be navigable under certain conditions" (Payer 1876, XLVI).

Conclusion

Let me return in conclusion to the questions I posed in the beginning: How was scientific knowledge communicated in various urban spaces, and how did it circulate between and within these spaces? As the various examples presented here have shown, knowledge about an ice-free ocean changed from context to context, sometimes being in the background as in the lecture hall in Vienna, or in the center as in that of Hamburg. Depending on the context, and on contemporary perceptions of facticity, such claims could be given different degrees of legitimacy. They might be presented as part of scientific discourse, as in Petermann's letters to the press in autumn 1871, or they might become part of popular discourse, as in the publication of the different opinions in the same media. What these examples also have shown is that it is futile to draw a clear line between these various forms of knowledge communication. Ultimately, these spaces were also media, that is they mediated knowledge and thus continuously transgressed the borders to other spaces. Knowledge circulation is always transgressive. Contemporary technologies of mobility and urban cultures of communication—telegraphy, steam travel, the press, maps, bodies, mass spectacles, and academic talks—helped define these spaces and the knowledge communicated there. They provided or removed evidence and legitimacy that was based on observations, eyewitness, and corporeal presence. The examples also suggest that the transfer between one medial form to another, and the different knowledge articulations within and across these media, could form more easily in the complex metropolitan space of Vienna than in the more restricted space of smaller cities such as Hamburg. At the same time, the two meetings in the Geographical Societies suggest that the transfer of scientific knowledge was easier in the Northern city of Hamburg, where Arctic knowledge was part of the everyday, than in the Southern metropolis of Vienna, where the Arctic was far away and little known. The Arctic became part of Vienna through discussions of its nature, sometimes more so, when scientific theories were discussed, sometimes less, when the Arctic sea became a projection space for Austrian daring and civilization, as in the 1872 exhibition or in the newspaper reports following the 1871 expedition. In all these cases, the global space of the Arctic was folded into Vienna, yet different epistemologies contributed to different forms of reception. This shows that geographical location is only one aspect that is relevant to understanding knowledge production and circulation.

Acknowledgement

I would like to thank Johan Schimanski and Mitchell G. Ash for their insightful comments on this chapter, and Kati Straner and Markian Prokopovych on an earlier version.

Bibliography

Abel, H., and Jessen, H. (1954). *Kein Weg durch das Packeis: Anfänge der deutschen Polarforschung (1868–1869)*. Bremen: Schünemann.

Ash, M. G. (2000). Räume des Wissens—was und wo sind sie? Einleitung in das Thema. *Berichte zur Wissenschaftsgeschichte* 23, 235–242.

Ash, M. G. (2006). Wissens- und Wissenschaftstransfer—Einführende Bemerkungen. *Berichte zur Wissenschaftsgeschichte* 29, 181–189.

Ash, M. G., and Surman, J. (2012). The Nationalization of Scientific Knowledge in Nineteenth-Century Central Europe: An Introduction. In: Ash, M. G., and Surman, J. (eds.), *The Nationalization of Scientific Knowledge in the Habsburg Empire, 1848–1918*. Basingstoke: Palgrave Macmillan, 1–29.

Aufruf für die Nordpolar-Expedition (1872). *Neue Freie Presse*, Morgenblatt, February 8.

Berger, F. (2015). *Julius Payer. Die unerforschte Welt der Berge und des Eises*. Innsbruck and Vienna: Tyrolia.

Berger, F., Besser, B. P., and Krause, R. A. (eds.) (2008). *Carl Weyprecht (1838–1881). Seeheld, Polarforscher, Geophysiker. Wissenschaftlicher und privater Briefwechsel des österreichischen Marineoffiziers zur Begründung der internationalen Polarforschung*. Vienna: Verlag der Österreichischen Akademie der Wissenschaften.

Bravo, M. T. (2019). *North Pole: Nature and Culture*. London: Reaktion.

Buchanan, I. (2018). Blue Humanities. In: *A Dictionary of Critical Theory* (2nd ed.). Oxford: Oxford University Press. DOI: 10.1093/acref/9780198794790.001.0001.

Chavanne, J. (1874a). Oesterreichs Nordpol-Expedition. *Neue Freie Presse*, Morgenblatt, September 19.

Chavanne, J. (1874b). Die Nordpolfrage und die Ergebnisse der zweiten Österr.-Ungar. Nordpolar-Expedition. *Petermann's Geographische Mittheilungen* 11, 421–425.

Daum, A. (2002). *Wissenschaftspopularisierung im 19. Jahrhundert: Bürgerliche Kultur, naturwissenschaftliche Bildung und die deutsche Öffentlichkeit, 1848–1914* (2nd rev. ed.). Munich: Oldenbourg.

Der Nordpol (1874). *Illustrirtes Wiener Extrablatt*, October 1.

Die Ausstellung der Nordpol-Expedition (1872). *Gemeinde-Zeitung*, April 19.

Die Ausstellung der österreichischen Nordpol-Expedition in Wien (1872). *Illustrirtes Wiener Extrablatt*, April 23.

Die Nordpolfahrer (1874). *Die Presse*, Abendblatt, September 25.

Die Nordpolfahrer in der Geographischen Gesellschaft (1874). *Neue Freie Presse*, Morgenblatt, September 30.

Die Nordpolfahrer in Hamburg (1874). *Neue Freie Presse*, Morgenblatt, September 24.

Expedition nach dem Nordpolarmeer (1872). *Neue Freie Presse*, Abendblatt, February 7.

Felsch, P. (2010). *Wie August Petermann den Nordpol erfand*. Munich: Luchterhand Literaturverlag.

Finnegan, D. A. (2011). Placing Science in an Age of Oratory: Spaces of Scientific Speech in Mid-Victorian Edinburgh. In: Livingstone, D. N., and Withers, C. W. J. (eds.), *Geographies of Nineteenth-Century Science*. Chicago and London: University of Chicago Press, 153–177.

Franz Josefs-Land! (1874). *Illustrirtes Wiener Extrablatt*, September 5.

Hellwald, F. v. (1874). Die österreichisch-ungarische Nordpol-Expedition und die Nordpol-Frage. II. *Wiener Abendpost*, October 19, 1908.

Holtorf, C. (2012). Die Südsee im Norden. Technik und Transzendenz in Narrativen der Arktisforschung des 19. Jahrhunderts. In: Neumeister, K., Renger-Berka, P., and Schwarke, C. (eds.), *Technik und Transzendenz. Zum Verhältnis von Technik, Religion und Gesellschaft.* Stuttgart: W. Kohlhammer, 181–208.

Holtorf, C. (2014). Das offene Polarmeer. Ein Bilddiskurs im 19. Jahrhundert. In: Eder, F. X., Kühschelm, O., and Linsboth, C. (eds.), *Bilder in historischen Diskursen.* Wiesbaden: Springer Fachmedien, 145–172.

Holtorf, C. (2017). Zum Pol. Positionen der Arktis in Raum und Zeit. In: Meyer, H., Rau, S., and Waldner, K. (eds.), *SpaceTime of the Imperial.* Berlin: De Gruyter Oldenbourg, 220–244.

Howes, H. S. (2013). *The Race Question in Oceania: A. B. Meyer and Otto Finsch Between Metropolitan Theory and Field Experience, 1865–1914.* Frankfurt am Main: Peter Lang.

In der Geographischen Gesellschaft (1874). *Die Presse,* Morgenblatt, September 30.

Keskitalo, E. C. H. (2002). *Constructing the "Arctic": Discourses of International Region-Building.* Rovaniemi: Lapin yliopisto.

Koldewey, C. (1871). Eisverhältnisse im grönländischen Meere und Ansichten über weitere Förderung arktischer Entdeckungen. *Hansa: Zeitschrift für Seewesen* 10: 89–92.

Koselleck, R. (1979). *Vergangene Zukunft. Zur Semantik geschichtlicher Zeiten.* Frankfurt am Main: Suhrkamp.

Krause, R. A. (2010). *Daten statt Sensationen. Der Weg zur internationalen Polarforschung aus einer deutschen Perspektive* (= Berichte zur Polar- und Meeresforschung 609). Bremerhaven: Helmholtz Gemeinschaft. hdl:10013/epic.34343

Lefebvre, H. (1992). *The Production of Space,* ed. D. Nicholson-Smith. Malden: Blackwell.

Lightman, B. (2007). *Victorian Popularizers of Science: Designing Nature for New Audiences.* Chicago: University of Chicago Press.

Littrow, [Heinrich von]. Die Nordpolfahrer (1874). *Die Presse,* Abendblatt, September 24.

Livingstone, D. N., and Withers, C. W. J. (eds.) (2011). *Geographies of Nineteenth-Century Science.* Chicago: University of Chicago Press.

Luedtke, B. (2015). An Ice-Free Arctic Ocean: History, Science, and Scepticism. *Polar Record* 2, 130–139. https://doi.org/10.1017/S0032247413000636.

Murphy, D. T. (2002). *German Exploration of the Polar World: a History, 1870–1940.* Lincoln, NE: University of Nebraska Press.

Oberlieutenant Payer über die Entdeckung des offenen Polarmeeres (1871). *Fremden-Blatt,* Morgen-Blatt, November 30.

Oesterreichische Nordpolfahrer (1871). *Wiener Zeitung,* October 5.

Östling, J., Heidenblad, D. L., Sandmo, E., Hammar, A. N., and Nordberg, K. (2018). The History of Knowledge and the Circulation of Knowledge: An Introduction. In: Östling, J., Sandmo, E., Heidenblad, D. L., Hammar, A. N., and Nordberg, K. (eds.), *Circulation of Knowledge: Explorations in the History of Knowledge.* Lund: Nordic Academic Press, 9–33.

Payer, J. (1871). Schreiben des Herrn Jul. Payer über die Entdeckung eines offenen Polarmeeres an den Verein für Geographie and Statistik zu Frankfurt a. M. *Gaea. Natur und Leben. Zeitschrift zur Verbreitung naturwissenschaftlicher und geographischer Kenntnisse sowie der Fortschritte auf dem Gebiete der gesammten Naturwissenschaften* 7: 630–632.

Payer, J. (1874a). Die Nordpol-Expedition. Dampfer 'Finnmarken', im September 1874. *Neue Freie Presse*, Morgenblatt, September 25.

Payer, J. (1874b). Die zweite Osterr.-Ungar. Nordpolar-Expedition unter Weyprecht und Payer, 1872/4: K. K. Ober-Lieutenant J. Payer's offizieller Bericht an das Comité, d. d. September 1874. *Petermann's Geographische Mittheilungen* 12, 443–451.

Payer, J. (1876). *Die österreichisch-ungarische Nordpol-Expedition in den Jahren 1872–1874, nebst einer Skizze der zweiten deutschen Nordpol-Expedition 1869–1870 und der Polar-Expedition von 1871*. Vienna: Alfred Hölder.

Petermann, A. (1871a). Die Entdeckung eines offenen Polarmeeres durch Payer und Weyprecht im September 1871. *Wiener Zeitung*, October 13.

Petermann, A. (1871b). Die Entdeckung eines offenen Polarmeeres durch Payer und Weyprecht im September 1871. *Petermann's Geographische Mittheilungen* 11: 423–424.

Potter, R. A. (2007). *Arctic Spectacles: The Frozen North in Visual Culture, 1818–1875*. Seattle and London: University of Washington Press.

Rechnungsabschluß des Comité für die österreichisch-ungarische Nordpol-Expedition (1874). *Wiener Zeitung*, December 18.

Sarasin, P. (2011). Was ist Wissensgeschichte? *Internationales Archiv für Sozialgeschichte der deutschen Literatur (IASL)* 1, 159–172. https://doi.org/10.1515/iasl.2011.010.

Schimanski, J., and Spring, U. (2009). Explorers' Bodies in Arctic Mediascapes: Celebrating the Return of the Austro-Hungarian Polar Expedition in 1874. *Acta Borealia* 16: 50–76.

Schimanski, J., and Spring, U. (2015). *Passagiere des Eises. Polarhelden und arktische Diskurse 1874*. Vienna: Böhlau Verlag.

Singer, F. I. (1874). *Unsere Nordpolfahrer. Ein Wort an das Volk zur Feier der Rettung und Heimkehr der Helden des Tages*. Vienna: Buchhandlung von F. I. Singer.

Spring, U. (2018). Die Arktis als Wiener Wissensraum. Öffentlichkeit und Wissenschaft im späten 19. Jahrhundert. In: Feichtinger, J., Klemun, M., Surman, J., and Svatek, P. (eds), *Wandlungen und Brüche. Wissenschaftsgeschichte als politische Geschichte*. Vienna: V&R unipress, Vienna University Press, 311–316.

Spring, U., and Schimanski, J. (2015). The Useless Arctic: Exploiting Nature in the Arctic in the 1870s. *Nordlit* 35, 13–27.

Spring, U., and Schimanski, J. (2019). The Melting Archive: The Arctic and the Archives' Others. In: Frank, S. K., and Jakobsen, K. (eds.), *Arctic Archives: Ice, Memory, and Entropy*. Bielefeld: Transcript, 49–68.

Treibhölzer aus dem nördlichen Eismeere (1872). *Neue Freie Presse*, Abendblatt, February 15.

Von der Nordpol-Expedition (1874). *Die Presse*, Morgenblatt, September 24.

Von der Polar-Expedition der Herren Payer und Weyprecht (1871). *Wiener Abend-post*, October 27.

Walsh, S. A. (2016). Liberalism at High Latitudes: The Politics of Polar Exploration in the Habsburg Monarchy. *Austrian History Yearbook* 47, 89–106. http://doi.org/10.1017/S0067237816000084.

Werner, M., and Zimmermann, B. (2006). Beyond Comparison: *Histoire Croisée* and the Challenge of Reflexivity. *History and Theory* 45:1, 30–50.

Weyprecht, C. (1874). Die zweite Österr.-Ungar. Nordpolar-Expedition unter Weyprecht und Payer, 1872/4: K. K. Schiffs-Lieut. C. Weyprecht's offizieller Bericht an das Comité, d. d. /2. Sept. 1874. *Petermann's Geographische Mittheilungen* 11: 417–421.

Withers, C. W. J., and Livingstone, D. N. (2011). Thinking Geographically about Nineteenth-Century Science. In: Livingstone, D. N., and Withers, C. W. J. (eds.), *Geographies of Nineteenth-Century Science*. Chicago: University of Chicago Press, 1–19.

Zukünftige arktische Reisen (1874). *Hansa: Zeitschrift für Seewesen* 11: 21, 161–163.

7 Academic Geography and its Networks in Vienna and Berlin

A First Comparative Study

Petra Svatek

Introduction

As Jürgen Osterhammel has already noted for the German-speaking countries, in multiple fields a "multiplication and acceleration" of "interactions took place especially across national borders" during the nineteenth century; above all the second half of the century can be described as "a period of unprecedented networking" (Osterhammel 2009, 1010–1011). This thesis also applies to geographical research in Vienna from the 1880s onwards. According to social theorist David Harvey, this is the first time in the history of Viennese geography that an acceleration in research can be ascertained by "overcoming spatial barriers" (Harvey 1995, 345).

After the University reform led by Leopold Graf von Thun Hohenstein (1811–1888), a professorship for geography was established at the University of Vienna and occupied by the natural scientist Friedrich Simony (1813–1896) in 1851. Because geographical research also began in other scientific disciplines, geography in Vienna had become an indispensable field from the 1880s onward. Not only geographers, but also botanists, physicians, geologists, ethnologists, archaeologists, and historians repeatedly carried out geographically oriented research. The two major research centers were the University and the Imperial and Royal Academy of Sciences. Various networks emerged, mainly within Vienna, but also within the Habsburg Monarchy and across national borders. Geoscientific collective projects, in which the metropolis of Vienna acted as a hub within the Habsburg Monarchy, existed at that time not only in geography but also, for example, in the fields of geology and meteorology. Major examples are the Geological Survey of Austria, which was carried out by the Imperial and Royal Geological Reichsanstalt, founded in 1849 (Klemun 2012), as well as work on the geographical location of earthquakes (Coen 2012) and studies of the climate of the Habsburg Monarchy, all organized by the Central Institute for Meteorology and Geomagnetism, founded in 1851 (Coen 2018).

In contrast to Vienna, the "densification" (Harvey 1995, 345) of geographical research and networks began in Berlin soon after the Friedrich

Wilhelm University was founded in 1810. In the first half of the nineteenth century, Carl Ritter (1779–1859), in particular, gave the Geographical Department an internationally influential impetus. However, other scientists, such as Alexander von Humboldt (1769–1859), and institutions such as the Gesellschaft für Erdkunde zu Berlin, founded in 1828, also devoted themselves to geographical research (Lenz 2003; Schultz 2010).

During their research, Viennese and Berlin scientists produced knowledge about the planet Earth as a human habitat and provided facts with texts, maps, and other forms of representation, which in turn served as the basis for new interpretations of geographical knowledge. They examined the Earth's physical environment and the effects of its forces and processes (for example in ice age and climate research and plant geography) as well as cultural-geographical themes such as the dynamics and structure of society, economic activity, and ethnically and culturally constructed "spaces" studied under the heading of "cultural morphology" (www.spektrum.de/lexikon/geographie/geographie/2917. Download 31 October 2018).

For reasons of space, this chapter discusses only the Geographical Departments at the Universities of Vienna and Berlin from the 1880s to the 1920s. Within this time frame, Friedrich Simony, Albrecht Penck (1858–1945), Wilhelm Tomaschek (1841–1901), Eugen Oberhummer (1859–1944), Eduard Brückner (1862–1927), Philipp Paulitschke (1854–1899), Erwin Hanslik (1880–1940), Hugo Hassinger (1877–1952), and Otto Lehmann (1884–1941) were the main contributors at the University of Vienna, while Ferdinand von Richthofen (1833–1905), Heinrich Kiepert (1818–1899), Fritz Jäger (1881–1966), Albert Herrmann (1886–1945), and from 1906 also Albrecht Penck were employed at the University of Berlin.

Since no other books and articles have yet been published on this topic, and since extensive archival studies would have to be carried out in order to do justice to the subject, only a preliminary overview of the networks established at these Geographical Departments can be given here (for broader surveys see Lichtenberger 2001 and Schultz 2010). The aim of this chapter is to work out the similarities and differences between Vienna and Berlin on the basis of selected criteria such as internationalization and politicization. A central common denominator of these two research departments is the fact that both of them, due to their locations in the capitals of the two Central European empires, functioned primarily as nodes of research networks. Geographers in these two cities were hardly involved in research or other activities that were not organized from Vienna or Berlin. These functions of supra-regional decision-making, innovation, and competition undoubtedly also favored the development of geographical science in these places. In this chapter, the question of how geographers networked with political authorities and other scientists will be examined in more detail. Both in Vienna and in Berlin, networks were established within the metropolis, with other cities, and also with other nations. The geographers at these two locations were thus integrated into networks of different research groups, institutions, and resources. But there are differences between Vienna and Berlin with respect to the international scope of cooperation and the political orientation of their respective research.

1. The Department of Geography at the University of Vienna

When Simony retired from the University of Vienna in 1885, the professorship of geography was divided into two positions, a physiogeographical and a cultural-geographical branch, and the Geographical Department was formally established; as a result, the number of networks that emanated from this department increased. However, with only a few exceptions (such as the "Atlas of the Austrian Alpine Lakes" discussed next), these networks rarely resulted in joint publications. The contacts seem to have been used primarily for the acquisition of data and for the mutual exchange of knowledge.

In the second half of the 1880s, the most extensive joint project initiated solely by the Geographical Department was the "Atlas of the Austrian Alpine Lakes", which resulted in cooperation within the department, with other Viennese institutions, and also outside Vienna. This project connected the Vienna center with peripheral alpine regions, but showed hardly any internationality or politicization. The working group consisted of Friedrich Simony, Albrecht Penck, and Johann Müllner (1869–1952) from the Geographical Department at the University of Vienna and Eduard Richter (1847–1905) from the Geographical Department of the University of Graz. In addition, project head Albrecht Penck cooperated with the Viennese publisher Eduard Hölzel, who printed the atlas in two volumes in 1895/1896 (Penck and Richter 1895, 1896). The project was influenced by Simony, who had already carried out soundings in the Salzkammergut lakes on his own initiative during the 1840s. Simony developed a specially constructed sounding apparatus, which he combined with a depth thermometer and let down into the water innumerable times, measuring the distances by counting strokes of the boat's oars. This gave him important data on the shape of previously unexplored lake basins (Svatek 2015a, 51–52; Kretschmer 1996, 46–47; Kainrath 2013, 305–316). Simony drew a first depth chart of Lake Hallstatt in 1845. Perhaps he had assistants, who are not yet known. Furthermore, he could already refer to international experience, collected for example in the depth measurements of Swiss lakes. As early as 1819 the English geologist Henry Thomas de la Bèche (1796–1855) had carried out depth measurements in Lake Geneva. Around 1840, the Swiss naturalists Arnold Guyot (1808–1884) and Louis-François de Pourtalès (1823–1880) measured Lake Neuchâtel and Lake Morat (Pestalozzi 1894, 59–60). Whether Simony cultivated personal contacts with these researchers or merely drew upon the published literature could not be determined.

This project is a good example of how the network of Viennese and Graz geographers spread Simony's thinking and research practices and integrated this into the work of other scientists. An "exchange of ideas or a mental interaction" of the kind described by Ludwik Fleck (1935, 45–46) was indispensable for this project and also for all other projects in the following years. The scientists carried out depth measurements according to Simony's specifications, but a division of labor can be observed. While Simony continued to research the depths of the Salzkammergut lakes, Richter was responsible for the Carinthian and South Tyrolean lakes. The depth charts were produced by Simony and Müllner, while Penck and Richter wrote the accompanying texts. Looking at these works, a hierarchical structure can also be discerned. Müllner may only have acted according to the specifications of the other geographers, which suggests that they placed him at the lower end of the hierarchy. Penck, who held the chair of physiogeography at the University of Vienna until 1906, was the lead supervisor of the project. The Imperial and Royal Ministry for Religious Affairs and Education provided the financing, which also established a connection to politics. However, the initiative for this financing seems to have come primarily from the Hölzel publishing house and not from the geographers or the ministry. Nor did the atlas have any special economic significance. The political significance of the project lay in the fact that the Austrian Alps were already central symbols for the Habsburgs and specifically Austrian patriotism. This was also the reason why the project focused only on the Austrian and not the Swiss and Bavarian lakes.

The "Atlas of the Austrian Alpine Lakes" was the only project during the period under review in which an intensive cooperation among the geographers of the Viennese Department and with geographers from other Austrian universities could develop. In the next few years, further partnerships followed, such as the research carried out by Penck and Brückner on the glaciation of the Alps and the Alpine foothills during the Ice Age (The Alps in the Ice Age 1909). Penck, who had already carried out geological mapping in southern Bavaria since 1880 (Henniges 2017b, 322–411), was able to draw on preliminary work in this field. For Penck and Brückner, "mapping in the field and the evaluation of specialist literature" were "an elementary way of producing, consolidating and changing thoughts" (Henniges 2017b, 394). Countless discussions took place between the two geographers regarding data evaluation and interpretation, while they mapped the terrain separately. In this project, areas outside the Habsburg Empire (Switzerland and the Bavarian foothills of the Alps) were also integrated into the research.

From the mid-1880s onwards, regions outside Europe came into focus of geographers for the first time. Privatdozent Philipp Paulitschke traveled to the Horn of Africa in 1884/1885 together with the Moravian physician Dominik Kammel Edler von Hardegger (1844–1915). The resulting publications, which were based on Paulitschke's research alone, were the first ones on the ethnographic and geological conditions in Northeast Africa (Paulitschke 1891). Paulitschke also contacted other travelers to Africa, notably Count Eduard Wickenburg (1866–1936) and Ludwig von Höhnel (1857–1942), and produced route maps and accompanying texts according to their data (Kretschmer 1988, 152; Zeilinger 1999, 210–218). Paulitschke's contacts were mainly with researchers and explorers of the Habsburg Monarchy. His networks therefore had a certain regional orientation. A similar finding applies to Wilhelm Tomaschek, who researched areas outside Europe and their history, languages, and historical landscapes, but did not form extensive networks with scientists outside the Habsburg Empire. According to the current state of research, he seems to have obtained his knowledge less through direct contacts with scientists than by "condensing and copying paperwork" from libraries and archives (Friedrich and Zedelmaier 2017, 273).

Many of the complex interactions at the Geographical Department took place through correspondence, then as now "the decisive platform for research, thinking and argumentation" in academic science (Assmann 2010, 300). During the 1890s, Albrecht Penck was able to build up the most extensive networks of scientists inside and outside geography and the most diverse arrangements at the Department of Geography of the University of Vienna. Letters in the Archive of Geography at the Leibniz Institute for Regional Geography tell us that Penck not only worked with all the leading geographers of the German Reich and Austria-Hungary (Alfred Hettner, Eduard Richter, Robert Sieger, Alfred Kirchhoff, Joseph Partsch, Friedrich Ratzel, Hermann Wagner, etc.) between 1897 and 1899, but also with scientists of related disciplines, such as the geologist Rudolf Hoernes, the meteorologist Julius von Hann, the anthropologist Felix von Luschan, and the mineralogist Theodor Liebisch. Outside the German-speaking countries he corresponded, for example, with the French geographer Emmanuel de Martonne, the Scottish geographer John Scott Keltie, the American geographers William A. Libbey (Princeton University), Richard E. Dodge (Columbia University), and William Morris Davis (Harvard University) as well as with the Dutch geologist Gustaaf Adolf Frederik Molengraaff, the English geologist Thomas Mellard Reade, and representatives of various institutions (the Société de Géographie de Genève, the American Museum of Natural History New York, the United States Geological Survey, and the Iowa Geological Survey).[1] Which information was exchanged and how Penck made use of it for his own research need to be examined in more detail.

Of the other Viennese geographers, Eugen Oberhummer seems to have cultivated the most numerous contacts beyond national borders up to the 1920s. He succeeded Tomaschek in 1903, but had already established a network with many scientists working in the fields of historical geography, art history, and archaeology (such as Max Ohnefalsch-Richter, Heinrich Schliemann, Fritz von Miller, and Georg Schweinfurth) during his associate professorship at the University of Munich from 1892 to 1902 (Oberhummer 1890, 183–240; Oberhummer 1892, 420–486). His travels alone testify to extensive acquaintances, which Oberhummer regarded mainly as data suppliers; he published largely alone. During the First World War, for example, he traveled several times to Southeastern Europe, from where he obtained and published information about administrative changes and topographic surveys in areas occupied by the military that were hardly accessible to scientists in Vienna (Oberhummer 1918, 313–346; Oberhummer 1922, 116–117. For excerpts from Oberhummer's travel diaries see Bertele and Wacker 2004).

During his travels, Oberhummer was integrated into an extensive network of scientists, teachers, lawyers, and military personnel. One example is his 1917 trip to Montenegro and Albania, organized by the Imperial-Royal War Press Headquarters and led by dragoon Lieutenant Willy Dittrich. Representatives of various professions took part, such as the lawyer Robert Bartsch (1874–1955), the economist Josef Gruntzel (1866–1934), the lawyer Rudolf Pollak (1864–1939), school principal Gustav Krützner, and chamber of commerce consultant Emil Perels (Oberhummer 1918, 313–332). Oberhummer maintained local contacts above all with representatives of the state administration and the military, who also willingly provided him with information containing militarily sensitive data, such as strategic maps. Thus he was allowed to inspect some sheets of the map of Montenegro, which at that time was not accessible to the public and was only produced for political use (Oberhummer 1918, 335–336).

In 1904, 1910, and 1912, Oberhummer traveled to the United States, Canada, and Mexico, where he lectured at the University of Chicago, Yale University, the University of California at Berkeley, Harvard University, and John Hopkins University, among others, and made contact with geographers such as William Morris Davis at Harvard.[2] These relationships eventually helped Oberhummer to gain a special honor in 1926/1927, his participation in the so-called Swimming University. In this undertaking, more than 450 students from many American universities were shipped around the world for a full academic year while maintaining their studies. Oberhummer was the only foreigner on the ship among the 50 lecturers. He gave lectures in English on historical geography and on the geography of the countries being visited, such as Cuba, Japan, the Philippines, Thailand, Indonesia, and India (on this journey see Oberhummer 1926). For Oberhummer, the trip also represented mutual networking, since he not only made his knowledge available to the students and other participants, but also had the opportunity to conduct his own research and gain knowledge about other countries.

He reported, for example, that through personal contacts with the King of Siam he was able to view official data that were hardly accessible to foreigners (Oberhummer 1929, 346). The data obtained during this trip were incorporated into publications on Siam, Shanghai, and Japan (Oberhummer 1929; Oberhummer 1932a; Oberhummer 1932b; Oberhummer 1934).

From the beginning of the twentieth century onwards, the assistants and lecturers of the Geographical Department did not act as a group, but often acted on their own initiative and maintained contact with only a few cooperation partners within the Habsburg Empire or Vienna. In the first half of the 1910s, Hugo Hassinger mainly dealt with the planned structural redesign of the Viennese city area and mapped all buildings according to their date of construction. He understood his research as a resource for more efficient urban planning, since all buildings that had been built from the Middle Ages to the 1840s and still existed were to be included in the redesign of the cityscape and not demolished (Hassinger 1912; Hassinger 1916). If the age of a house could not be identified exactly, he searched in Viennese archives for sources that gave him information about the building's architectural history. Hassinger was thus one of the first Viennese geographers to combine field and archival research (Svatek 2019, 127–128).

In this project Hassinger was probably involved in an exchange of ideas with only two people. Franz Wilhelm Englmann (1862–1926), curator of the Historical Museum of the City of Vienna, helped him study old city plans. From Emil Tranquillini (1884–1955), professor at the Technical University of Brno, he received ground plans and elevations of old Viennese houses (Dvořák 1916, I; Hassinger 1916, 6). Finally, Hassinger's research helped him to establish new acquaintances in the Viennese monument and heritage protection movement, which gave him the opportunity to publish his study in the book series "Österreichische Kunsttopographie" edited by Viennese monument protector Karl Giannoni (1867–1951) and Max Dvořák (1874–1921), Professor of Art History at the University of Vienna. The funds required for this work were provided by the Vienna City Council (Dvořák 1916, I). However, it has not yet been possible to determine whether Hassinger's research actually contributed to more efficient Viennese urban planning.

Until the 1920s, Otto Lehmann's networks hardly extended beyond national borders. Although he had also studied in Leipzig and Paris, the acquaintances he made during his residency in Vienna were apparently of no significance. His study of a landslide at the Sandling Mountain near Leoben in Styria (1920) was commissioned by Oberhummer, who was able to raise the needed funds as a member of the Historical-Philosophical Class at the Austrian Academy of Sciences. Lehmann seems to have carried out the research alone, although at about the same time Fritz Machatschek (1876–1957) from the Charles University in Prague and Erich

Spengler (1886–1962) from the Federal Geological Survey also investigated this natural disaster. That no mutual exchange of ideas took place between Lehmann and the other gentlemen is suggested by the fact that they published different research results concerning the demolition of the rock masses (Lehmann 1926).[3]

In another project, the exploration of the "ice giant world" (located inside the Hochkogel Mountain near Werfen) initiated by the Austrian Academy of Sciences, Lehmann worked in a group consisting of various Viennese scientists and Salzburg speleologists. These included the biologist Otto Wettstein (1892–1967), the chemist Ernst Hauser (1896–1956), and the geologist Julius Pia (1887–1943). Lehmann seems to have worked in partnership with Pia to investigate the geomorphological conditions of the ice cave, while the other scientists mainly investigated its meteorological and biological conditions (Mattes 2016, 177–178). Lehmann was also involved in a working group on the standardization of signs in cave maps. Together with the cartographers Karl Peucker (1859–1940) and Ludwig Teißl, the geologist Gustav Götzinger (1880–1969), and the speleologist Rudolf Saar (1886–1963) and under the chairmanship of the geodesist and university professor at the Vienna School of Technology Eduard Dolezal (1862–1955), he created a cave sign key, which was published in 1925 (Mattes 2015, 261).

In contrast to Lehmann, Erwin Hanslik was not only able to establish his networks within the scientific community, but had also extensive contacts with artists. In 1915 he founded the Vienna Department for Cultural Research together with the Viennese Orientalist Edmund Küttler (1884–1964). This department brought together members from the fields of art and culture, such as Gustav Klimt (1862–1918), Otto Wagner (1841–1918), and Adolf Loos (1870–1933). The aim was to analyze the cultures of the earth objectively and make a contribution to "applied war research" (Feichtinger 2013, 116; Hanslik 1916). In this context Hanslik developed an extremely bizarre world view, which was rejected by most geographers. He located Europe in the south up to the region where "the closed tropical plant cover begins" (Hanslik 1917, 87), so that North Africa was also assigned to the European cultural area. In the East, Hanslik regarded the "Himalayas, Tienshan and Altai" as a border, since "a coherent plant cover" would characterize this "earth space" and could thus "separate the European natural and human kingdom from the rest of the earth" (Hanslik 1917, 87. For further details see Henniges 2015, 1332–1335; Schultz 2012, 40–41; Svatek 2015b). Much more fruitful scientifically, however, was Hanslik's morphological research on cultural boundaries and cycles (Hanslik 1907), which became an important source for Penck's theory of German "folk and cultural soil" (*Volks- und Kulturboden*) (Henniges 2015, 1316–1323, 1336–1340). Here we have an example of knowledge transfer between Vienna and Berlin that came about as a result of Penck's migration to Berlin.

2. The Geography Department of the Friedrich Wilhelm University of Berlin

In comparison to Vienna, the Geography Department at the University of Berlin saw a more intensive internationalization and a somewhat different kind of politicization in this period. While there was hardly any cooperation within the department, almost all Berlin geographers had extensive networks beyond Central Europe and traveled to non-European regions. Transnational connections were established, resulting in transfer of knowledge across national borders, which was hardly the case among Viennese geographers in the nineteenth and early twentieth centuries. In addition, Berlin geographers cooperated directly with political authorities and received commissions from them, and their research had an impact on politics at the time.

When Ferdinand von Richthofen assumed the professorship of Physical Geography in Berlin in 1886, he was already an internationally renowned scientist, above all because of his geological, economic and geographical research on China. He stayed in contact with many scientists, such as the Swedish geographer and explorer Sven Hedin (1865–1952). China's economically important regions were to come under German influence at this time, and Richthofen's "reference to the Shantung peninsula was included in the decision-making process for the acquisition of Kiautschous" (Schultz 2010, 664). Although a personal intervention by Richthofen has not yet been proven, his knowledge of Kiautschou Bay was undoubtedly important for politicians during the negotiation of its lease to the German Reich. Kaiser Wilhelm II himself, Admiral Alfred Peter Friedrich Tirpitz (1849–1930) of the Imperial Naval Office, the diplomat Gustav Adolf Schenck zu Schweinsberg (1843–1909), German envoy to China from 1893 to 1896, and the Asia expert and diplomat Maximilian August Scipio von Brandt (1835–1920), Schweinsberg's predecessor in China, studied Richthofen's work (Wardenga 1990, 151).

Heinrich Kiepert probably maintained the most intensive contact with explorers. Like Paulitschke, he also produced route sketches and maps of previously little explored areas, for example for Gustav Nachtigal (1834–1885). He also made maps for scientists from other disciplines. Otto Benndorf (1838–1907), who was appointed to the chair of archaeology at the University of Vienna in 1877 as Alexander Conze's successor (1831–1914), received from him an overview map for his work "Reisen in Lykien und Karien" (1884). Kiepert also created maps for the archaeologists Christian Hülsen (1858–1935) and Robert Koldewey (1855–1925), who worked at the German Archaeological Department in Rome (Zögner 1999, 25, 127, 128, 130).

During his professorship in Berlin, Albrecht Penck was without doubt "the central hub of German-speaking and international geography" (Henniges 2017a, 571). He cultivated a particularly large number of contacts with US scientists, which he established during his visiting professorships at Columbia University (1908/1909), Yale University (1908/1909), and the University of California at Berkeley (1928). In 1927, for example, Penck was invited by the American Philosophical Society to give a lecture in Philadelphia. The Society was then celebrating its 200th anniversary, for which he conveyed "congratulations from the Berlin and Vienna Academy of Sciences, as well as the University of Berlin".[4] In this case, Penck also acted as a representative of the Academy of Sciences in Vienna. While in the US he took the opportunity to travel to the Southwest, accompanied by his private assistant Albrecht Haushofer (1903–1945). Penck made contacts with German consulates and American geographers (such as the previously mentioned William Morris Davis), who helped him with the technical execution of the trip. On behalf of the Reich Ministry of Food and Agriculture, he also visited the First International Soil Science Congress in Washington, DC.[5]

Penck was the Berlin geographer who cultivated the most intensive contacts with politics and published the most politically significant publications in his field until the 1920s. Immediately after the end of the First World War, Penck initiated several cartography-related projects, which were intended to provide information for the Paris peace negotiations about the ethnographic conditions in Poland and in the eastern part of the Reich. The map showing the "Distribution of the Germans and Poles in West Prussia and Poznan",[6] produced by Penck in cooperation with the cartographer Herbert Heyde, should be cited here as an example. However, Penck's maps were not considered in the German preparations for the Paris peace negotiations (Herb 1997, 22–28; Pinwinkler 2011, 187). His previously mentioned study on German "folk and cultural soil", an initial "ideological consolation ['weltanschaulicher Trost'] that was supposed to help recover from the insult to Germany at Versailles" (Schultz 2018, 122), later achieved considerably more political influence. After the National Socialists took power, this "folk and cultural soil" ideology served the new regime as a "national geography" (ibid., 123). The extent to which Penck's racist theory actually became the basis of Nazi occupation policy is still debated today (Haar 2008; Schultz 2018).

Fritz Jäger, who was appointed to the newly founded chair of colonial geography in Berlin in 1911, also placed his work at the service of politics. In this function he undertook several trips to Africa and Latin America. Having already taken part in an expedition to German East Africa in 1904 under the geographer Carl Uhlig (1872–1938) (Jäger 1905), he carried out his own studies there from 1914 on the orders of the Reich Colonial Office. This research was clearly suggested to him by politicians. His work mainly included topographical, plant-geographical, morphological, and geological surveys, which he carried out with the geographer Leo Waibel (1888–1951); they thus provided important regional data on this German colony (Michaelsen 1914).

Jäger's research in Mexico in 1925 was also not initiated by him, but carried out in connection with a trip by German industrialists. He and the other participants were invited on this trip by the Mexican government "in order to strengthen the economic and cultural relations between Mexico and the German Reich" (Jäger 1926, VII). Jäger extended his stay to carry out physiogeographical research. The trip was financed by the Prussian Ministry of Science, Art and Public Education, the Foreign Office, and the Emergency Committee for German Science (*Notgemeinschaft der Deutschen Wissenschaft*). Jäger established extensive networks with Mexican authorities and scientists. He carried out his own research in cooperation with the Mexican Ministry of Agriculture and Cultural Work, which granted the Tyrolean geologist of the Direccion de Estudios Geograficos y Climatológicos, Paul Waitz (1876–1961), a leave of absence to assist Jäger. The Mexican geographer José Luís Osorio Mondragón (1885–1944) from the University of Mexico and the Swiss geologist Arnold Heim (1882–1965), who was in Mexico at the time, also accompanied him on many excursions. Jäger also received extensive scientific information from the geologist Ernst Wittich (1871–1952), from the Mexican Society for Geography and Statistics, and other sources (Jäger 1926, VII–VIII).[7]

This journey did not have any colonial scientific aspect, like those undertaken before 1918. But even in the 1920s, Berlin geographers were involved in a venture that made at least a certain colonial claim: the "German Atlantic Expedition" from 1925 to 1927, which explored the depths and water conditions of the southern Atlantic Ocean between Africa and South America in 14 profiles. Under scientific objectives "the German presence in waters visible to the whole world" could be legitimized, which "remained closed to the navy" after the First World War (Höhler 2002, 239). Initiated by Friedrich Schmidt-Ott (1860–1956) and financed by the Prussian Ministry of Culture and the Emergency Committee for German Science, this trip resulted in networking among geographers, oceanographers, mineralogists, biologists, meteorologists, geologists, and chemists. The "German Atlantic Expedition" is probably the most comprehensive interdisciplinary collaboration of the two geographical departments in the period under study. In addition to Fritz Haber (1868–1934) from the Kaiser Wilhelm Institute for Physical Chemistry and Electrochemistry, the chemist Hermann Wattenberg (1901–1944) from the Technical University of Gdansk, the mineralogist Carl Wilhelm Correns (1893–1980) from the Prussian Geological Survey, and the geologist Otto Pratje (1890–1952) from the German Hydrographic Department were also involved. In addition, Alfred Merz (1880–1925), Georg Wüst (1890–1977), and Albert Defant (1884–1974) from the Department of Oceanography, which at the time was run in personal union with the Geographical Department of the Friedrich Wilhelm University, participated (Höhler 2002, 234–246; Hoheisel-Huxmann 2007).

Albert Herrmann also built up an extensive international network during his research into the historical geography of Chott el-Djerid in southern Tunisia in the 1920s. He was advised by experts from various countries and institutions: the ancient historian Pere Bosch i Gimpera (1891–1974) from the University of Barcelona, the ancient philologist Edgar Martini (1871–1932) from the German University of Prague, the archaeologist Aurel Stein (1862–1943) from the Archaeological Survey of India, the papyrologist Friedrich Bilabel (1888–1945) from the University of Heidelberg, the archaeologist August Albert von Le Coq (1860–1930) from the Berlin Ethnological Museum, and the Orientalist Hans von Mžik (1876–1961) from the Austrian National Library. Eugen Oberhummer also appeared in the list, which suggests cooperation between Vienna and Berlin.[8] However, we do not yet know about the intensity of the exchange of knowledge or the information exchanged. Herrmann eventually located the mythical empire "Atlantis" in Chott el-Djerid (Herrmann 1930).

Conclusion

As this chapter has shown, the networks of geographers at the Universities of Vienna and Berlin were both local, regional, and supraregional. Both an exchange of ideas and the exchange of information and maps with other persons or institutions were involved in these networking activities. At the two geographical departments, networking became more and more part of scientific life; without such activities, data could not be obtained and travel could not be carried out or financed.

However, there were differences between Vienna and Berlin in the intensity of internationalization and politicization. The geographers of the Friedrich Wilhelm University in Berlin included non-European areas such as China, Mexico, German Southwest Africa, and elsewhere in their research, but also cooperated with many scientists from different countries. The extent to which these relationships were actually reciprocal remains to be investigated. Did Berlin and Viennese geographers gain more knowledge and information from other scientists than they themselves made available to others? How did the knowledge transfer process function in this context? Such knowledge transfers were often only partial (Renn and Hyman 2012, 21), for example in the case of Fritz Jäger and his research in Mexico, whose handwritten notes contained far more information than he actually published.[9]

Berlin geographers were engaged in extensive networking with policymakers and political authorities during the period under study. They received commissions from the political side (for example Fritz Jäger from the Reich Colonial Office) and made research available to political authorities on their own initiative (as did Albrecht Penck in the course of the Paris peace negotiations); thus it cannot always be assumed that the sciences were somehow "recruited" by political actors (Ash 2010, 17). In contrast, Viennese geographers neither received commissions directly from politicians and political authorities nor does their work seem to have been used by politicians as a basis for decision-making. But geographers did act politically, and their research had also a political background. This is clear in the case of Hassinger's study on Viennese urban planning and Oberhummer's treatises on political geography, mentioned previously. A change occurred after Hugo Hassinger's appointment to the chair of cultural geography and the founding of the Southeast German Research Association in 1931. From this time onward Viennese geographers increasingly placed themselves at the service of politics and worked on research projects that already had a connection to National Socialist Germany before 1938, such as the Burgenland Atlas (Svatek 2017; Svatek 2018; Svatek 2019). Later, both Viennese and Berlin geography played important roles during the Nazi era, providing research results either on their own initiative or at the request of political authorities for the scientific preparation of the reorganization of the German living space and Nazi occupation policy.

Notes

1. The letters are archived in the Archive for Geography, Leibniz-Institut für Länderkunde Leipzig IfLA, estate Albrecht Penck 871/7–9.
2. From October 6 to November 3, 1910, for example, Oberhummer gave a cycle of 21 lectures entitled "The Political Geography of Europe" at the Department of Geography at the University of Chicago. At the Washington Academy of Sciences he spoke on December 5, 1910, about "The Races and Peoples of Europe". See the announcement sheets of the University of Chicago and the Washington Academy of Sciences (Archive of the Austrian Academy of Sciences, estate Oberhummer 103.017).
3. While Machatschek and Spengler assumed a "slippage of the rock masses on a slippery base" (Lehmann 1926, 164), Lehmann interpreted the landslide as "a deposit of firm, fissure-like-permeable rocks over a strongly soaked clayey base" (Lehmann 1926, 299).
4. Archive of the Berlin-Brandenburg Academy of Sciences and Humanities ABBAW, PAW (1812–1945), II-VII-165 (Albrecht Penck, Bericht über meine Reise nach Nordamerika 14.04.-10.08.1927).
5. For a report on the trip see ibid.
6. The map was published in: Zeitschrift der Gesellschaft für Erdkunde zu Berlin (1919), map 1.
7. See also IfLA, estate Fritz Jäger, 849 and 850 (Diaries and geographical observations Mexico).
8. Secret State Archives Prussian Cultural Heritage Foundation GStA PK, I. HA Rep. 76 Ministry of Education, Va Sekt. 2, Tit. IV, Nb. 55, Vol. 8.
9. His written records are archived in the IfLA, estate Fritz Jäger.

Bibliography

Archival Sources

Archive of the Berlin-Brandenburg Academy of Sciences and Humanities (ABBAW): PAW (1812–1945), II-VII-165.
Archive for Geography, Leibniz-Institut für Länderkunde Leipzig (IfLA): estate Fritz Jäger 849 and 850; estate Albrecht Penck 871/7–9.
Secret State Archives Prussian Cultural Heritage Foundation (GStA PK): I. HA Rep. 76 Ministry of Education, Va Sekt. Two, Tit. IV, Nb. 55, Vol. 8.

Primary and Secondary Literature

Ash, M. G. (2006). Wissens- und Wissenschaftstransfer—Einführende Bemerkungen. *Berichte zur Wissenschaftsgeschichte* 29, 181–189.
Ash, M. G. (2010). Wissenschaft und Politik. Eine Beziehungsgeschichte im 20. Jahrhundert. *Archiv für Sozialgeschichte 50*, 11–46.
Assmann, A. (2010). Kooperieren und Korrespondieren vom Briefwechsel zum Email Exerzitium. In: Hennig, J., and Andraschke, U. (eds.), *Weltwissen: 300 Jahre Wissenschaften in Berlin*. Munich: Hirmer, 300–302.
Bertele, M., and Wacker, C. (2004), *Die Reisetagebücher Eugen Oberhummers. Die Reisen in die Alte Welt*. Munich: Oberhummer Gesellschaft e. V.
Coen, D. R. (2012). Fault Lines and Borderlands: Earthquake Science in Imperial Austria. In: Ash, M. G., and Surman, J. (eds.), *The Nationalization of Scientific Knowledge in the Habsburg Empire, 1848–1918*. Basingstoke: Palgrave Macmillan, 157–182.
Coen, D. R. (2018). *Climate in Motion. Science, Empire, and the Problem of the Scale*. Chicago: University of Chicago Press.
Dvořák, M. (1916). Vorwort. In: *Kunsthistorischer Atlas der k. k. Reichshaupt- und Residenzstadt Wien und Verzeichnis der erhaltenswerten historischen Kunst- und Naturdenkmale des Wiener Stadtbildes*. Vienna: Schroll, i.
Feichtinger, J. (2013). Kulturwissenschaften.at. Varianten, Traditionen und Entwicklungen in Österreich. Ein Essay. In: Höhne, St. (ed.), *Kulturwissenschaft(en) im europäischen Kontext. Fachhistorische Entwicklungen zwischen Theoriebildung und Anwendungsorientierung*. Frankfurt am Main: Lang, 109–122.
Fleck, L. (1935). *Entstehung und Entwicklung einer wissenschaftlichen Tatsache. Einführung in die Lehre vom Denkstil und Denkkollektiv*. Basel: Schwabe.
Friedrich, M., and Zedelmaier, H. (2017). Bibliothek und Archiv. In: Sommer, M., Müller-Wille, S., and Reinhardt, C. (eds.), *Handbuch Wissenschaftsgeschichte*. Stuttgart: Metzler, 265–275.
Haar, I. (2008). Leipziger Stiftung für deutsche Volks- und Kulturbodenforschung. In: Fahlbusch, M., and Haar, I. (eds.), *Handbuch der völkischen Wissenschaften*. Munich: Saur, 374–382.
Hanslik, E. (1907). *Kulturgrenze und Kulturzyklus in den polnischen Westbeskiden: eine prinzipielle kulturgeographische Untersuchung*. Gotha: Perthes.
Hanslik, E. (1916). *Das Institut für Kulturforschung*. Wien: Institut für Kulturforschung.
Hanslik, E. (1917). *Österreich. Erde und Geist*. Vienna: Institut für Kulturforschung.

Harvey, D. (1995). Zeit und Raum im Projekt der Aufklärung. *Österreichische Zeitschrift für Geschichtswissenschaften* 6:3, 345–365.

Hassinger, H. (1912). *Wiener Heimatschutz- und Verkehrsfragen.* Vienna: Freytag & Berndt.

Hassinger, H. (1916). *Kunsthistorischer Atlas der k. k. Reichshaupt- und Residenzstadt Wien und Verzeichnis der erhaltenswerten historischen Kunst- und Naturdenkmale des Wiener Stadtbildes.* Vienna: Schroll.

Henniges, N. (2015). "Naturgesetzte der Kultur": Die Wiener Geographen und die Ursprünge der "Volks- und Kulturbodentheorie". *ACME: An International E-Journal for Critical Geographies* 14:4, 1309–1351.

Henniges, N. (2017a). Albrecht Penck. In: Fahlbusch, M., Haar, I., and Pinwinkler, A. (eds.), *Handbuch der völkischen Wissenschaften. Forschungskonzepte— Institutionen—Organisationen—Zeitschriften.* Berlin: De Gruyter, 570–577.

Henniges, N. (2017b). *Die Spur des Eises. Eine praxeologische Studie über die wissenschaftlichen Anfänge des Geologen und Geographen Albrecht Penck (1858–1945).* Leipzig: Selbstverlag Leibniz-Institut für Länderkunde e.V.

Herb, G. H. (1997). *Under the Map of Germany: Nationalism and Propaganda 1918–1945.* London: Routledge.

Herrmann, A. (1930). Forschungen am Schott el-Djerid und ihre Bedeutung für Platons Atlantis. *Dr. A. Petermann's Mitteilungen aus Justus Perthes Geographischer Anstalt* 76, 243–250.

Herrmann, A. (1934). *Unsere Ahnen und Atlantis: nordische Seeherrschaft von Skandinavien bis nach Nordafrika.* Berlin: Klinkhardt & Biermann.

Hoheisel-Huxmann, R. (2007). *Die Deutsche Atlantische Expedition 1925–1927.* Hamburg: Convent-Verlag.

Höhler, S. (2002). Profilgewinn. Karten der Atlantischen Expedition (1925–1927) der Notgemeinschaft der Deutschen Wissenschaft. *NTM—Zeitschrift für Geschichte der Wissenschaften, Technik und Medizin* 10, 234–246.

Jäger, F. (1905). Bericht über den Anfang der deutsch-ostafrikanischen Expedition der Otto Winter-Stiftung unter Leitung von Prof. Dr. C. Uhlig. *Zeitschrift der Gesellschaft für Erdkunde zu Berlin* 40, 215–217.

Jäger, F. (1914). Die Forschungsreise von Prof. F. Jäger nach Deutsch-Südwestafrika. *Zeitschrift der Gesellschaft für Erdkunde zu Berlin* 49, 569–573.

Jäger, F. (1926). *Forschungen über das diluviale Klima in Mexiko.* Gotha: Perthes.

Kainrath, W. (2013). "Johann Oskar" Friedrich Simony zum 200. Geburtstag. *Mitteilungen der Österreichischen Geographischen Gesellschaft* 155, 305–316.

Kater, M. (2006). *Das "Ahnenerbe" der SS 1935–1945. Ein Beitrag zur Kulturpolitik des Dritten Reiches.* Munich: Oldenbourg.

Klemun, M. (2012). National, Consensus' as Culture and Practice: The Geological Survey in Vienna and the Habsburg Empire (1849–1867). In: Ash, M. G., and Surman, J. (eds.), *The Nationalization of Scientific Knowledge in the Habsburg Empire, 1848–1918.* Basingstoke: Palgrave Macmillan, 83–101.

Kretschmer, I. (1988). Österreichs Beitrag zur kartographischen Erschließung Ostafrikas bis zum Ersten Weltkrieg. In: *Abenteuer Ostafrika. Der Anteil Österreich-Ungarns an der Erforschung Ostafrikas. Ausstellungskatalog.* Eisenstadt: Amt d. Bgld. Landesregierung, 129–160.

Kretschmer, I. (1996). Kartographische Arbeiten Friedrich Simonys. *Geographischer Jahresbericht aus Österreich* 53, 43–61.

Lehmann, O. (1926). *Die Verheerungen in der Sandlinggruppe (Salzkammergut) durch die im Frühherbst 1920 entfesselten Naturgewalten.* Vienna: Hölder-Pichler-Tempsky.

Lenz, K. (2003). Erneuerung durch Wandel: Entwicklungsperioden in der Geschichte der Gesellschaft für Erdkunde. *Die Erde. Zeitschrift der Gesellschaft für Erdkunde zu Berlin,* 7–16.

Lichtenberger, E. (2001). Geographie. In: Acham, K. (ed.), *Geschichte der österreichischen Humanwissenschaften* (Vol. 2). Vienna: Passagen, 75–148.

Mattes, J. (2015). Underground Fieldwork: A Cultural and Social History of Cave Cartography and Surveying Instruments in the 19th and at the Beginning of the 20th Century. *International Journal of Speleology* 44:3, 251–266.

Mattes, J. (2016). Going Deeper Underground: Social Cooperation in Early Twentieth-Century Cave Expeditions. In: Klemun, M., and Spring, U. (eds.), *Expeditions as Experiments: Practising Observation and Documentation.* London: Palgrave Macmillan, 163–186.

Michaelsen, W. (1914). Eine Forschungsreise von Prof. Dr. Fritz Jäger nach Deutsch-Süd-West-Afrika. *Zeitschrift der Gesellschaft für Erdkunde zu Berlin* 49, 153–154.

Oberhummer, E. (1890). Aus Cypern: Tagebuchblätter und Studien. *Zeitschrift der Gesellschaft für Erdkunde zu Berlin* 25, 183–240.

Oberhummer, E. (1892). Aus Cypern: Tagebuchblätter und Studien 2. *Zeitschrift der Gesellschaft für Erdkunde zu Berlin* 27, 420–486.

Oberhummer, E. (1918). Montenegro und Albanien unter österr.-ungar. Verwaltung. *Mitteilungen der Geographischen Gesellschaft in Wien* 61, 313–346.

Oberhummer, E. (1922). Die Kriegsaufnahme von Albanien. *Petermanns Geographische Mitteilungen* 68, 116–117.

Oberhummer, E. (1926). Amerikanische Universitätsreise. *Neue Freie Presse,* Nr. 22372, 27 December, 1–2.

Oberhummer, E. (1929). Siam. Eindrücke und Studien. *Mitteilungen der Geographische Gesellschaft in Wien* 72, 346–376.

Oberhummer, E. (1932a). Schanghai. *Mitteilungen der Geographischen Gesellschaft in Wien* 75, 5–27.

Oberhummer, E. (1932b). Japans Häfen. *Mitteilungen der Geographischen Gesellschaft in Wien* 75, 300–302.

Oberhummer, E. (1934). Weltreligionen. In: Haushofer, K. (ed.), *Raumüberwindende Mächte.* Leipzig: Teubner, 110–139.

Osterhammel, J. (2009). *Die Verwandlung der Welt. Eine Geschichte des 19. Jahrhunderts.* Munich: Beck.

Paulitschke, P. (1891). Uebersicht über die Völkerlagerung auf dem Osthorn von Afrika. Begleitwort zur Karte. *Mittheilungen der k. k. Geographischen Gesellschaft in Wien* 34, 468–475.

Penck, A., and Richter, E. (1895). *Atlas der österreichischen Alpenseen. Die Seen des Salzkammergutes.* Vienna: Ed. Hölzel.

Penck, A., and Richter, E. (1896). *Atlas der österreichischen Alpenseen. Seen von Kärnten, Krain und Südtirol.* Vienna: Ed. Hölzel.

Pestalozzi, S. (1894). Ueber Tiefenmessungen in schweiz. Seen. *Schweizerische Bauzeitung* 23/24–9.

Pinwinkler, A. (2011). "Hier war die große Kulturgrenze, die die deutschen Soldaten nur zu deutlich fühlten . . ." Albrecht Penck (1858–1945) und die deutsche "Volks- und Kulturbodenforschung". *Österreich in Geschichte und Literatur mit Geographie* 2, 180–191.

Renn, J., and Hyman, D. (2012). The Globalization of Knowledge in History: An Introduction. In: Renn, J. (ed.), *The Globalization of Knowledge in History*. Berlin: Edition Open Access.

Schultz, H.-D. (2010). Was "ist" Geographie? Was "ist" sie nicht? Zur Konfiguration des Faches als politisch relevante "reine" (Natur) Wissenschaft. In: Tenorth, H.-E. (eds.), *Geschichte der Universität unter den Linden 1810–2010, Band 5: Transformation der Wissensordnung*. Berlin: Akademie-Verlag, 651–674.

Schultz, H.-D. (2012). Europa, Russland und die Türkei in der "klassischen" deutschen Geographie. In: Reuber, P., Strüver, A., and Wolkersdorfer, G. (ed.), *Politische Geographien Europas—Annäherung an ein umstrittenes Konstrukt*. Berlin: LIT-Verlag, 25–54.

Schultz, H.-D. (2018). Albrecht Penck: Vorbereiter und Wegbereiter der NS-Lebensraumpolitik? *E&G Quaternary Science Journal* 66, 115–129.

Svatek, P. (2015a). "Natur und Geschichte": Die Wissenschaftsdisziplin "Geographische" und ihre Methoden an den Universitäten Wien, Graz und Innsbruck bis 1900. In: Ottner, C., Holzer, G., and Svatek, P. (eds.), *Wissenschaftliche Forschung in Österreich 1800–1900. Spezialisierung, Organisation, Praxis*. Göttingen: VR unipress, 45–71.

Svatek, P. (2015b). Geopolitische Kartographie in Österreich 1917–1937. *Mitteilungen der Österreichischen Geographischen Gesellschaft* 157, 301–322.

Svatek, P. (2017). Geographisches Institut der Universität Wien. In: Fahlbusch, M., Haar, I., and Pinwinkler, A. (eds.), *Handbuch der völkischen Wissenschaften. Forschungskonzepte—Institutionen—Organisationen—Zeitschriften*. Berlin: De Gruyter, 1398–1405.

Svatek, P. (2018). Kontinuität oder Wandlung? Stadt- und Landesplanung am Geographischen Institut der Universität Wien um 1938. *GW-Unterricht. Zeitschrift des Vereins für geographische und wirtschaftliche Bildung* 152:4, 5–13.

Svatek, P. (2019). Hugo Hassinger (1877–1952). Volkstumsforscher, Raumplaner, Kartograph und Historiker. In: Hruza, K. (ed.), *Österreichische Historiker. Lebensläufe und Karrieren 1900–1945* (Vol. 3). Vienna: Böhlau, 123–155.

Wardenga, U. (1990). Ferdinand von Richthofen als Erforscher Chinas. *Berichte zur Wissenschaftsgeschichte* 13:1, 141–155.

Zeilinger, E. (1999). Im Banne Afrikas. Philipp Paulitschke (1854–1899). Geograph, Ethnograph, Afrikareisender. Unpublished Ph.D. Dissertation, University of Vienna.

Zögner, L. (1999). *Antike Welten, neue Regionen: Heinrich Kiepert 1818–1899*. Berlin: Staatsbibliothek.

Web Sites

Blotevogel, H.-H. Geographie. www.spektrum.de/lexikon/geographie/geographie/2917. download 31 October 2018.

8 Capital Collections, Complex Systems

Vienna, Berlin, and Ethnographic Specimen Exchanges in Transnational *Fin de Siècle* Scientific Networks

Brooke Penaloza-Patzak

Introduction: Metropoles and Museum Aggregations

We tend to think of metropoles as bounded geographic spaces in which political and economic powers converge, resulting for a time in 'mother cities' that are the nexus of activity for, and administer to the needs of, their surrounding regions: Teotihuacan, Rome, London, and Paris all have been or still are metropoles in this sense. In the history of science, the word was used by Darwin to refer to areas within the "polity of nature" in which special differentiation came about as a result of site-specific confluences of climate, geography, and social organization (Darwin 1886, 93, 134–137). Expanding on both definitions, the following discussion will demonstrate how ethnographic collections in museums converged around the turn of the twentieth century with political and economic interests to co-create the distinct but interrelated practices that defined anthropology in Vienna and in Berlin, the cosmopolitan centers of the Austro-Hungarian and German Empires. This discussion will also delineate how collection exchange practices helped expand what had up until that point been the Europe-centered landscape of metropolitan science westward across the Atlantic. Home to the royal collections of their respective empires, Vienna and Berlin were also the geographic spaces within each empire in which a newly emerging species of scientist—the anthropologist (used here, unless otherwise indicated, to refer to individuals broadly involved in carving out what are today considered the fields of cultural and social anthropology, ethnology, and physical anthropology)—was most thickly concentrated.

Until 1866 the Austrian Empire, with Vienna as its seat, had enjoyed the status of a major power within Central Europe and the world more broadly. Following the Austro-Prussian War, however, Vienna began slowly ceding this status to Berlin, capital of the increasingly powerful Prussian Empire. Despite their relative geographic and linguistic propinquity, and the fact that each was the de facto nucleus of anthropology in its respective empire, there is little evidence of overlapping engagements between practitioners working in the two cities before the 1890s. This absence is all the more marked in light of a coincident increase of exchange with North American colleagues in both capitals, and this in spite of a general understanding that the Americans—with the possible exceptions of those at the Smithsonian Institution in Washington, DC, and the Peabody Museum in Cambridge, Massachusetts—were working at a material and analytical disadvantage.

This recent affinity was prompted by a growing interest in the Americas which was part of a broader European-wide scientific trend. During the late eighteenth and early nineteenth centuries, Captain James Cook's ethnographic collections and Alexander von Humboldt's reports from the new world had awakened among Europeans an intense interest in Native American cultures. When the Smithsonian Institution featured objects from Canada's Northwest Coast at the 1876 Centennial Exhibition in Philadelphia, it was thus only a matter of time before what Douglas Cole has memorably dubbed the "scramble for Northwest Coast artifacts" spread to Europe (Cole 1985, 48, 212, 222, 288; see also Penaloza-Patzak 2018c, 13, 33, 67 f. 138). This craze ignited Americanist anthropology west of the Atlantic, and presents an apt lens through which to examine the professional landscapes of late nineteenth-century anthropology, in particular collection and research practices, in Vienna and Berlin.

Merging considerations and methods from the history of science, social history, and museum anthropology, this comparative social history brings together and analyzes an internationally scattered body of correspondence and ethnographic specimens. Utilizing these sources, the chapter will present a multi-layered perspective that interweaves the scholars, collections, institutions, and ideas that populated and shaped the landscapes of anthropology in each of the two metropoles, as well as the transnational currents that mediated interaction and research between practitioners in these and other urban centers.

By 1883 Berlin's royal ethnographic department (at that point housed at the Neues Museum) had acquired one of the world's most extensive Pacific Northwest Coast collections, signaling the moment of its arrival within the transnational community of Americanist anthropologists. German anthropology was moving toward its zenith, and ingesting as much as possible of the wider world's material heritage. Austro-Hungarian, or more specifically Austrian anthropology, on the other hand, was turning inward, with an increased focus on the multi-ethnic populations within the borders of its own empire.

In Austria-Hungary and Germany as well as in the United States, the disciplinary framework of anthropology came together in the 1870s and 1880s in metropolitan museums, and slowly migrated into universities over the course of the late nineteenth and early twentieth centuries. *Fin de siècle* ethnographic collections in museums were thus the loci of anthropological knowledge as practiced and produced in those metropolitan centers. These were the spaces of and media for the scholarly transference of the practical skills, methods, and theoretical approaches that were coming to define, and eventually differentiate, the new field(s) of anthropology. Ethnographic collections are by nature relational structures. Driven by idiosyncratic professional considerations and curatorial approaches, they bring together and decontextualize cultural artifacts, introducing new associations between objects that often have little or no natural relation to one another. The processes by which this takes place are inherently embedded within the city in which the collection is located, its resources, needs, social and political climates; they are also impacted by the individuals who oversee and manage the collection, and their status within the national and international scientific communities. The metropole is in this sense the first unit of analysis and overarching structure that informs the shape and development of the scientific labor that takes place within it. Museums and their collections are nodes of activity within these macro structures, and exist in reciprocal and dynamic relation to the metropoles in which they came to reside.

Thus, the first order of business here will be to develop a comparative analysis of the Vienna and Berlin ethnological collections, snapshots situating each within the geography and culture of their respective cities, and to elucidate their roles, orientations, and reception within their national and international landscapes. In the second part of the chapter, a series of material exchanges carried out on the occasion of the 1893 World's Columbian Exposition in Chicago, Illinois, will serve as a means of elaborating on these analyses and bringing the professional landscape of anthropology in each metropole more sharply into focus. Characterizing metropolitan museums as 'thick spaces' within the 'complex system' of the transnational scientific landscape, and moreover spaces with the potential to become intertwined via specimen exchanges, we will see how the exposition acted upon the orientations and alignments of anthropology in Vienna, Berlin, and Chicago, inviting fresh points of contact, and providing the foundation of and impetus for new and more robust transfer networks (Brantz et al. 2012, 16–17; on complex systems and interactions among their constituent parts, see Holovatch et al. 2017, 2–3).

With a focus on transnational exchange, this research expands on earlier work by scholars such as Matti Bunzl and Glenn Penny on the intellectual history of German anthropology and the local and imperial contexts that shaped German ethnographic collections (Bunzl and Penny 2006; Penny 2002). The aim here is to reframe *fin de siècle* ethnographic collections as sites that communicated specific disciplinary approaches, and linked transnational urban centers via shared collection and research interests. The exposition exchanges will allow us to pinpoint and track critical moments in the development of material and theoretical exchange, and to assess the extent to which the enactment of those exchanges informed the positioning of the Vienna and Berlin collections and scholars in the broader transnational landscape. Drawing on the work of Gosden, Larson, and Petch, this approach presumes that museums are elaborate "aggregations" comprising myriad entangled structural units: humans, objects, ideas, and practices (Gosden et al. 2007, 1).

During the past three decades, scholarly interest in the roles that objects in museums—in particular their more theoretical dimensions—have played in the history and sociology of science has ripened into a line of inquiry that has been especially fruitful with regard to the natural sciences (Byrne et. al. 2011; te Heesen and Vöhringer 2014). Within that increasingly robust body of scholarship, however, the history of the social sciences remains comparatively under-analyzed, and the potential of collections-based research to provide insight into that history is largely untapped. As this chapter will demonstrate, the integration of archival and collections-based research provides a means of elaborating and interpreting the transfer processes and structural dynamics that drive scientific transformation.

1. Imperial Collections, Universal Visions

We now turn to the metropoles and their collections. The city now known as Vienna dates back to the Iron Age and, in large part thanks to its position along the Danube Canal, had established itself as a trade center as early as the Middle Ages. The history of Vienna as sometime capital of the Holy Roman Empire dates back to the late thirteenth century, when it was pressed into service as the seat of the Babenberg, and later the Habsburg, dynasties (Baedeker 1896, 12–13; Czeike 2004, 41–42, 119). Berlin, a much younger city, was just beginning to be settled in the early thirteenth century. In the late fifteenth century, it was chosen as the seat of the Brandenburg margraviate, and then the Hohenzollern dynasty, and named the capital of Prussia in 1701 (Baedeker 1897, 19–22). In 1861 Berlin incorporated a number of outlying suburbs and overtook Vienna as the most populous German-speaking city. By 1900 Vienna and Berlin, with circa 1,648,335 and 1,888,574 inhabitants respectively, had joined London, New York, Paris, and Tokyo as the world's most densely populated capital cities (Sedlaczek et al. 1902, 32; Hirschberg 1903, 4).

The Berlin ethnographic collection had been extracted from the royal cabinet of curiosity and made the subject of a dedicated ethnographic department in 1875; a similar process occurred in Vienna in 1876. In the years that followed, imposing buildings were constructed in each city's center to house and exhibit these collections. In 1886 the Berlin collection was transferred to a dedicated building: the Royal Ethnological Museum (*Königliches Museum für Völkerkunde*, KMV), a massive four-story, pentagon-shaped construction with circular vestibule in the fashionable Friedrichstadt neighborhood. In 1889 the Vienna collection gained six sprawling rooms on the ground floor of the Imperial and Royal Natural History Museum (*k. k. Naturhistorisches Hof-Museum*), a four-story historicist monument crowned by a 213-foot-tall copper cupola situated on the Ringstrasse that encircles Vienna's city center. The symbolic positioning of these collections in the administrative and intellectual hearts of these empires is unmistakable; it reinforced each city's claim to political and economic preeminence in its region and the scientific authority of those who worked most closely with those collections.

The accessibility of each collection reveals something more about their roles and function in metropolitan life: Admission to both was generally free, with the exception of Tuesdays in Vienna (when the cost was 1 florin, what one might expect to pay for a night in one of city's finer hotels). However, the Berlin museum was open 6 days a week, including a total of 12 hours on Saturdays and Sundays, while the Vienna collection was open just 4 days a week, and a total of 9 hours on weekends (on Vienna: Baedeker 1896, 10–11, 30–33; on Berlin: Baedeker 1897, 18–19, 60–63). Generous visiting hours rendered the Berlin collection, and thus to a certain extent developments in the field of anthropology, more accessible to a broader public than the Vienna collection, which catered to scholars, tourists, and the leisure class.

Both collections were displayed in large, iron-framed glass cases within which the objects were organized by geographic origin (Dorsey 1899, 471). The latter detail is of some significance, for while the rationale behind the geographic, rather than typological, approach to ethnographic display had become a norm in German-speaking lands, the practice was still far from the rule within the broader international landscape. A report by US anthropologist George Dorsey, who conducted a tour of Central European Museums in 1887, provides a keenly observed outside perspective on each department. Dorsey described the Vienna collection as the finest example of specimen installation in all of Central Europe, small but "scientifically classified, well arranged and cased, and carefully labeled", with notable material from South America and the Pacific Islands (Dorsey 1899, 468, 470–471). He described Berlin's collection, in contrast, as exhaustive and well-conceived but chaotic. The museum possessed more artifacts than any other (and perhaps any two) in the world, but failed to effectuate its otherwise "excellent" labeling and illustration concept. In his estimation, the collection's weakest region was North America, which only serves to underscore the overall superiority of Berlin's other aquisitions, for between 1881–1883 the museum had accessioned Adrian Jacobsen's extensive Pacific Northwest Coast collection. This comprised over 6,000 inventory numbers and was recognized as the finest of its kind outside the older collections in St. Petersburg and Madrid. Jacobsen's field notes and lack of scientific training were the cause of some derision among those responsible for cataloguing the specimens, but there was no doubting the cultural and economic value of the collection itself, which lent the museum and its affiliates a heightened status within Americanist anthropology circles (Penaloza-Patzak 2018c, 43–64; cf. Dorsey 1899, 468–469).

The number and training of those who managed the collections at each museum were roughly proportionate to their difference in size and level of international recognition. Franz Heger (1853–1931) had studied geology, minerology, paleontology, geodesy, and Oriental studies before receiving a curatorial appointment at the Natural History Museum in Vienna under his mentor Ferdinand von Hochstetter in 1882, and was named director of the ethnological collection in 1884 following Hochstetter's death (Fischer et al. 1976, 11; Heinrich 2006/2007, 116). Heger was the first person in Vienna to be employed as a full-time anthropologist. In 1892, Heger, who had carried out research in Greece and the Balkans, and extensive fieldwork in the Russian Empire, took on Indologist Michael Haberlandt as an assistant. Adolf Bastian (1826–1905), director of the Berlin museum and nearly 30 years Heger's senior, had traveled the world extensively. In 1866 he had become one of the first Germans to complete a Habilitation in ethnology, and by the 1880s was recognized as a founding figure in the field. He directed the Berlin museum with the help of several permanent assistants, one conservator, and a steady flow of temporary scientific staff. Both collections relied on a combination of government allocations and private philanthropy, but the Berlin museum had a financial advantage over Vienna. In 1881 Bastian had organized the *Ethnologisches Hilfscomité für Vermehrung der Ethnologischen Sammlungen* (Ethnological Committee to Aid the Expansion of the Ethnological Collections), a small but generous group of private patrons funded ethnographic expeditions that fed into the museum's collection (Westphal-Hellbusch 1973, 65–68).

Outside museum halls, anthropology in each city was structured by and organized in special-interest societies comprising a combination of specialists and enthusiasts: the *Berliner Gesellschaft für Anthropologie, Ethnologie und Urgeschichte* (BGAEU, Berlin Society for Anthropology, Ethnology and Prehistory), founded in 1869, and the *Anthropologische Gesellschaft in Wien* (AGW, Vienna Anthropological Society), founded in 1870. The Berlin society, founded in part by Bastian, helped establish and maintain a certain strength and uniformity of purpose across the different spheres of anthropology in Berlin, and had taken a fairly international approach to membership, and most certainly to research, from the outset (Penaloza-Patzak 2018c, 43–48). The collection and research activities of the Berlin museum and the BGAEU were tightly interwoven and, like the collection itself, extended well beyond Germany's colonial interests.

The disciplinary structure of anthropology in Vienna was quite different. Although Heger was a leading figure in the AGW throughout the late nineteenth century, the museum and the society shared no inherent relationship. The AGW had been founded by geologist Baron Ferdinand Leopold von Andrian-Werburg, who had developed an interest in ethnology while traveling in the eastern regions of the Habsburg Empire; as many scholars have since remarked, the society displayed an early and marked nationalist tendency in both research priorities and in its membership (Feest 1995, 121; Ranzmaier 2011, 3, 6–8; Penaloza-Patzak 2018c, 44). In 1877 the society, which had originally established its own ethnographic collection, donated its holdings to the NHM, and from that time until 1893 the two worked in closer collaboration, with AGW members often gifting expedition findings to the NHM. This arrangement served the ethnographic collection well, for the museum lacked the financial resources to make significant purchases on its own (Heinrich 1995/1996, 14–30).

In the earlier decades of the nineteenth century, the Austrian crown had financed the Austrian Brazil Expedition (1817–1835) and the *Novara* Expedition (1857–1859), which had resulted in some of the ethnographic department's most extensive collections. Following the 1867 establishment of the dual monarchy, however, research priorities began to shift, and by the final decades of the century, AGW members' research had become just as, if not more, concerned with the multi-ethnic population within the borders of their new empire as with those outside of it. One notable expression of that trend was the *Kronprinzenwerk*, a comprehensive 24-volume analysis of the land and people that constituted the monarchy, which was published between 1886 and 1902 (Erzherzog Rudolf 1886–1902). From an early point in his career, Heger had endeavored to maintain within the ethnographic collection an aspect of openness and flexibility in both research and collecting, but in the final decade of the nineteenth century this orientation brought the collection to something of an impasse with the AGW's focus on Austria-Hungary and the Balkans (Ranzmaier 2013, 73).

By this time the broader aims of anthropology as a field of study had begun settling into place, and the circumstances just described had begun to have a clear impact on the embeddedness of each museum and its affiliated practitioners within the international landscape. Berlin anthropology stood at the forefront of the international field. Its many representatives traveled widely, attending a broad range of international congresses and conducting research in other collections, and thereby served to spread and magnify the museum's reputation. Bastian remained general director of the museum, but increasingly ceded direction of individual departments to assistants Karl von den Steinen, Albert Grünwedel, Felix von Luschan, and Eduard Seler, regional specialists with extensive field experience.

In Vienna the ethnographic collection and the AGW pulled the field in different directions; Heger's tendency toward an international, universalist ethnography put him at odds even with his assistant Haberlandt, who by 1898 had co-founded another museum specifically dedicated to European ethnology. Heger's collection was the finest of its kind within the Austro-Hungarian Empire; however, dedicated and engaged a correspondent as he was, he had neither the financial means nor the professional influence to mobilize a more extensive research plan that would expand the collection's international profile.

This account thus far emphasizes some of the more fundamental incongruences between the organization and orientation of anthropology in the two imperial capitals of the German-speaking world. While each collection sat near the heart of its respective metropole, the Berlin collection was granted a dedicated multi-storied edifice, maintained a collection that was constantly supplemented by colonial acquisitions from abroad and, thanks to the financial support of a dedicated group of patrons, was able to commission additional expeditions and purchases as needed. Within just a few years of moving into its new building, the Berlin museum's holdings had overfilled their generous accommodations. The Vienna collection was less grandiose. With the majority of colleagues focused on more local phenomena and objects, the growth of the ethnographic collection became increasingly dependent on Heger's ability to negotiate purchases with the limited funds at his disposal or to realize exchanges with other collections. Neither of these strategies resulted in the acquisition of first-rate material. Lacking the infrastructure to command a larger building or staff, the Vienna collection would continue to be subsumed within the larger imperial and royal collection well into the twentieth century.

With regard to engaged professionals, societies, funding, and accessibility, then, the Berlin museum was thickly embedded in the local and international landscapes of anthropology by 1900, occupying a highly visible location within that quickly institutionalizing field of knowledge. Without the funds and momentum to command recognition, the Vienna collection was precariously situated, but Heger's exhibition and labeling had proven capable of exciting international scholarly interest in a collection that in itself was of little special import. In anthropology as in other collections-informed sciences, classificatory systems and modes are central elements in the creation of meaning from the objects at hand. The broader scientific value and appeal of metropolitan collections is thus borne out not only by the constituent accessions, but also by their meanings as configured by the methods of processing and organization that enact each curator's world view and approach to the study of culture.

In that sense, the ethnographic collections in these museums were distillations of the material, social, and intellectual components that constituted anthropology in each metropole. As we will see, those material components could be mobilized in the interest of initiating or strengthening trans-metropolitan networks. Drawing on a series of exchanges initiated on the eve of the 1893 World's Columbian Exposition, the following case studies will delineate how individual interests in Vienna and Berlin brought specific ethnographic objects and practices into wider circulation, and how those circulations in turn informed broader disciplinary practices and transnational museum relations.

2.   Transnational Dispersions, Part 1: Exchanges for the Fair

In the late summer of 1894, nine large wooden crates addressed to Heger arrived at the NHM. They had traveled to Vienna via first train, then ship, and then train again, from the Field Columbian Museum in Chicago. Six of the crates contained segments of what, once assembled, would amount to a massive *papier mâché* cast (to be discussed in further detail alongside others in the following pages). The remaining three crates conveyed a more motley assortment of objects: two bundles of circa 300 arrows from the Solomon Islands and a motley assortment of around 160 ethnographic objects—baskets, tools, and a few masks—from the Pacific Northwest of Canada.[1] At about the same time, another set of four crates addressed to Bastian in Berlin arrived at the KMV from the same source. Packed within lay more *papier mâché* casts.[2] The exchange of ethnographic materials between practitioners working in established ethnographic collections was becoming increasingly more commonplace, but the Field Columbian Museum, which had just opened earlier that summer, was far from established. How did this fledgling institution come to ship so much material a quarter of the way around the world? And what can these transactions tell us about *fin de siècle* anthropology within and between Vienna, Berlin, and the United States?

The material within the crates had been on exhibit at the World's Columbian Exposition in Chicago from May to October 1893. Nearly razed by the Great Chicago Fire of 1871, the city had "risen Phoenix-like from its ashes" by 1890, its population exceeding the million mark and earning it the distinction of the US's "second city" (Pierce 1957, 20). That same year, city lobbyists went up against representatives from New York, Washington, DC, and St. Louis and secured the distinction of host of the World's Fair celebrating the 400th anniversary of Columbus's so-called discovery of the new world (Pierce 1957, 501–502; Hinsley and Wilcox 2016, xvii). The crates were packed and dispatched to Europe by William H. Holmes, the nascent museum's new curator of ethnology, as payment in kind for objects that the Vienna and Berlin museums had sent to Chicago to supplement the Ethnographic Department exhibitions organized for the fair. The exposition has recently been described as something of a watershed moment in the development of US anthropology, however, there has been little research into its role in the development of international relations in the field (Hinsley 2016b, 99). What is known thus far can be put briefly as follows: Frederic Ward Putnam, director of the Peabody Museum of Archeology and Ethnology, was appointed curator and chief of the Department of Ethnology for the World's Columbian Exposition in 1891. His plan was to implement a sweeping series of exhibitions that traced the "material and moral progress of American civilization" from the arrival of Columbus in the New World up to the present. This was his opportunity to represent US anthropology to an international audience, and in order to do so effectively he began culling ethnological material from all over the world, commissioning over 100 fieldworkers in North America alone, and pressing into service some 56 of the Smithsonian Institution's foreign correspondents (Hinsley 2016a, 15–22).

Shortly after his appointment, Putnam hired German-born anthropologist Franz Boas as his first assistant. Boas, who had trained under Bastian, Rudolf Virchow, and the KMV directorial assistants in the Berlin collection, moved to the US in 1886 in hopes of securing a permanent position. Since that time, he had occasionally served as an exchange agent, arranging ethnographic transfers between US and German museums (Penaloza-Patzak 2018a). For the purposes of the Chicago exposition, Boas had been engaged to oversee the Department of Ethnology, Physical Anthropology Division and soon initiated a side project, soliciting contributions from Bastian in Berlin and Heger in Vienna to supplement Putnam's limited Polynesian, South American, and African material with so-called ethnographic duplicates from those areas. Duplicates provide a fascinating example of mutable value construction in ethnographic collections (Nichols 2018).

At that time, the majority of ethnographic collections were divided into a three-tier hierarchy: the exhibition collection, the study collection, and duplicates. Duplicates were designated as such based on their perceived inferiority to extant type specimens within their home collection but, depending on that collection's overall quality, might be far superior to material in other collections. The perceived value of duplicates lay in the impression that they could be deaccessioned without posing harm to the representative integrity of the home collection and functioned as tender for the acquisition of more desirable specimens from exchange partners (Penaloza-Patzak 2018a, 31). Heger's collection was noted for its excellent Polynesian and Brazilian material acquired from the late eighteenth and early nineteenth century Cook and Natterer collections. Berlin had also acquired Polynesian material from the Cook expeditions, and its African collection had grown so extensive by the late 1880s that, as one museum curator put it, the constant accumulation of treasures had become a "nightmare" for all but specialists and the most serious students (Krieger 1973, 106).

Alongside his requests for ethnographic material, Boas also made an appeal for photographs, schematic diagrams, and narrative descriptions detailing the museums' floor plans, collection processing methods, and exhibition strategies, which would be used to construct a comparative and didactic presentation on museum practices within the ethnographic exhibition (Figure 8.1).[3]

These requests underscore the multiple levels of exchange to which the fair gave rise, as well as the potential for object-based research to aid in the reconstruction of moments of theoretical exchange in the history of the sciences. In images and words, through the use of detailed pictures and wall labels, little evidence of which remains today, Boas elucidated for specialists and the US fair-going public alike the theoretical principles underlying the geographic arrangement of ethnographic collections, an experience that was reproduced within the Berlin museum's ethnographic display. By 1893 geographically organized ethnographic exhibitions were not wholly foreign to the US context, but the practice was far from as widespread or systematically institutionalized as was the case in the German and Austro-Hungarian Empires. Isolated instances could be seen at prominent institutions such as the Smithsonian's United States National Museum, but the theoretical underpinnings of these continued to be

Figure 8.1 World's Columbian Exposition of 1893. Photographs, museum publications, and schematic diagrams detailing the Berlin and Vienna museums' floor plans and exhibition strategies (Berlin museum display at left, Vienna at rear). Image: Gift of Frederic Ward Putnam, 1893.

Source: Courtesy of the Peabody Museum of Archaeology and Ethnology, Harvard University, PM93–1–10/100266.1.26.

informed first and foremost by a deterministic evolutionary perspective (Jacknis 1985, 77–83). With regard to viewer experience the difference between the messages received, even on a subconscious level, from typological versus geographic arrangements cannot be overestimated. The former presented object types in sequential displays ordered according to technological complexity in such a way as to convey a unilinear conception of cultural evolution. The latter contextualized objects alongside others from the same geographic area, oftentimes incorporating maps, photographs, and other visual information that provided additional insight into the natural conditions which helped shape life in that area, spatially and visually activating the exhibition space in such a way as to suggest a plurality of cultures and prompt viewers to consider each object within its particular contexts. Boas's guiding intention here was to substantiate and contextualize his own preference for geographic exhibition, aquired in the course his training in the KMV collection, and which he had championed in the US since his arrival in 1886 (Penaloza-Patzak 2018a; Penaloza-Patzak 2018c, 103–118).

Heger agreed to provide both ethnographic and didactic materials from the Vienna collection, as long as the transaction was conceived as an exchange. He was particularly intent on obtaining material from Canada's Pacific Northwest. Considering the intended scope of Putnam's exhibits, it was clear that at the close of the exposition Chicago would be awash in ethnographic material from all over the world, which meant that Heger could leverage an exchange to acquire new North American material at little expense. An agreement with the Berlin museum, which possessed more independent means of growing its collection, was less immediately forthcoming. Bastian was away wrapping up a two-year expedition, and the request was passed from one to another of his assistants until it was eventually taken up by Eduard Seler.[4] As an Americanist who specialized in Mexico, Seler had something to gain from involvement with the exposition, both in terms of material collections and with respect to his professional network. In the end, Seler arranged for the exchange model that had been proposed by Heger to be adopted by Berlin. His readiness to take on Boas's requests was clearly fueled by the exposition's focus on the Americas, and his own desire to acquire a representative collection of ancient material from the Mississippi Valley.[5]

The fact that both Heger and Seler contributed to the exposition based on the prospect of receiving American material in return, along with Putnam's decision to focus his fair exhibits on the Americas, emphasizes the potential for even a temporary modification in the cosmos of a particular scientific landscape to harness the energies of previously more or less autonomous actors. This focus on American ethnographic material initiated a shared scientific endeavor, and therewith new exchange infrastructures, subtly reconfiguring the metropolitan ethnographic museum constellation. Assuming the exchange correspondence from Boas, Putnam cautioned Heger and Seler that he could not guarantee specific returns, but assured both that they would be fairly compensated. Broadly speaking, the understanding was that, following the close of the fair, each museum's contribution would remain in Chicago and be repaid with a combination of duplicates and cast reproductions culled from the materials brought together for the exposition. The Berlin collection was to benefit in particular as a result of Boas's embeddedness in that museum landscape and ongoing engagement with its collection. Bastian and Seler easily secured Boas's promise that he would personally oversee the installation of their ethnographic collection in accordance with written specifications regarding labeling and geographic organization, and would take care to assure that they received quality material in exchange for their contributions.[6]

Within weeks Heger, a timely and thorough correspondent, had prepared and addressed to Putnam in Chicago eight crates packed with over 1,000 objects—a general collection from Central Africa, an extensive collection of weapons and ornaments from Polynesia, and musical instruments from the Sunda Islands—plans and depictions of the Vienna museum and his exhibits, and seven volumes of the museum's annals, together valued at around $750.[7] A few months later Seler sent Berlin's more modest contribution, valued at roughly $200 and comprising 35 reproduction casts from Guatemala, maps and photographs of the Berlin museum and its exhibits, and a set of museum publications (*Report of the Massachusetts Board of World's Fair Managers* 1894, 163).[8] That Seler continued to address his exchange correspondence and shipments to Boas, rather than Putnam, underscores the fact that this was only one in a series of exchanges in which Boas served as a representative of the Berlin museum's US interests; in this respect he was Seler's natural point of contact. Heger's position was more peripheral within that German-speaking exchange network, and by extension Boas's. This was particularly the case with respect to Americanist anthropology, and would become increasingly evident in the course of events leading up to the conclusion of the exchange.

Once complete, the Department of Ethnology Building at the World's Columbian Exposition housed a total of 314 different exhibits. The Vienna and Berlin museums' plans and illustrations were displayed along a 22-foot-long by 10-foot-high wall and accompanied by their respective publications (as previously shown in Figure 8.1).[9] According to contemporary reports, the African and Polynesian material Heger had sent was considered "one of the most valuable European collections" on display, and the casts from Seler, shown in a roughly 5-by-10 square meter space that reproduced Berlin's geographically organized exhibition schema, helped provide fairgoers with a singular opportunity to study the "remains of the cultured prehistoric races of North America" (*Rand, McNally & Co.'s A Week at the Fair* 1893, 105; Johnson 1898 vol. 3, 421, 424–426).

3. Transnational Dispersions, Part 2: Exchanges From the Fair

Aware that he might not be in a position to settle the exchanges once the exposition was over, Boas set about securing what material he could, namely plaster cast reproductions, while the fair was still underway. The second story of an unused dairy barn, referred to as the "Anthropology Annex", had been given over to this purpose, and before decamping from Chicago, Boas commissioned a team of cast-makers to produce material for his exchanges (Figure 8.2).[10]

Having made no specific promises to Heger, Boas ordered a *papier mâché* reproduction of an ancient Mayan idol from Quiriguá, Guatemala, that would account for one-half of the Vienna exchange and left the rest of Heger's exchange to be sorted out after the fair.[11] No such chances were taken with the Berlin exchanges. Boas instructed the cast makers to reproduce 16 small sculptures from Copán, Honduras; one cast of a tattooed head; two Mexican boomerangs; and a small altar, altar panel, idol, and set of reliefs that, like the idol for Vienna, had originated from

Figure 8.2 World's Columbian Exposition, 1893. Plaster cast technician crew with reproduction Yucatan ruins in a former dairy barn repurposed for use as the "Anthropology Annex". Image: Gift of the American Antiquarian Society, 1959.

Source: Courtesy of the Peabody Museum of Archaeology and Ethnology, Harvard University, PM59–50–10/100265.1.2.

Quiriguá (Johnson 1898 vol. 2, 316; and *Rand, McNally & Co.'s A Week at the Fair* 1893, 106).[12]

The idea of repaying the exchanges with casts, more accurately casts of casts, was pragmatic. The Central American exhibits were a popular success, had garnered much attention in contemporary accounts of the fair, and were widely reproduced in photographs and illustrations (Johnson 1898 vol. 2, 316; *Rand, McNally & Co.'s A Week at the Fair* 1893, 106; *Glimpses of the World's Fair* 1893, u. p.). Prominent among these had been what appeared to be stone monoliths, heads and stelae from Copan and Quiriguá, but were in fact *papier mâché* casts of these objects that had been produced between 1891 and 1892 by fieldworkers working in Central America working under Putnam's direction. Once dried, the casts were transported to Chicago, carefully reconstructed on site and painted to resemble the originals (*Report of the Massachusetts Board of World's Fair Managers*, 162). Comprising alternate layers of paper and plaster, the casts of those casts that Boas commissioned could be reproduced *ad infinitum* at a relatively low cost. While this rendered them materially inferior to ethnographic duplicates, the research value they posed for Americanist colleagues working in Europe more than compensated for that deficiency.

Just before the end of the fair, Chicago department store magnate Marshall Field agreed to finance a new natural history museum to house the material shipped to the city for the exposition, which is known today as the Field Museum. The fair had made good on its "promise of progress through science and technology", laying the groundwork for the future of anthropology in Chicago and earning the city the recognition of Europe's "increasingly global . . . metropolitan intellectual system and its scientific elite" (Wilcox 2016, 125; Hinsley 2016a, 55–59). Boas, who was hired to stay on after the fair and help organize the collections in preparation for their transfer to the new museum, entertained intermittent hopes for an appointment as curator of anthropology. Had this materialized, he would have been well positioned to pursue additional exchanges with the Vienna and Berlin museums. As it was, however, the appointment, along with responsibility for the outstanding portions of the Vienna exchange, went to the aforementioned William H. Holmes, previously of the Smithsonian Institution's Bureau of American Ethnology. Advised of Heger's request before Boas's departure, the majority of objects Holmes selected to make up for the remaining exchange value were indeed from Canada's Pacific Northwest, but the materials chosen conveyed a certain lack of investment in the transaction. Rather than selecting a few representative pieces, Holmes sent 172 lesser-quality duplicates generally varying in value from around 50¢ to $5, with two pieces—a model house and crest pole set (cat. no. 51.775, 51775a) and a wooden bird mask (cat. no. 51.747)—each valued at $25. Accompanied by vague records of provenance, the objects were described according to form or function rather than cultural relevance, thus rendering their scientific potential negligible, especially for a collection such as Heger's, which was known for its meticulously systematic displays.

Once the exchanges were complete, it appeared that the Berlin museum, with its strong ties to Boas that occasioned special attention with regard to the cast exchanges, had come out ahead of the Vienna museum, which was primarily repaid with ethnographic duplicates, at least in terms of the research potential of the exchanges. As with any history involving museum collections, however, this one did not end with the 1894 accession of the objects. Over the years Vienna's exposition duplicates have been shifted from storage area to storage area, never quite garnering the attention necessary to make it onto the exhibition floor. New strands of research indicate, however, that at least some of the pieces Holmes sent in the Vienna assemblage are indeed traceable, and thus possess more recognizable cultural and research value than initially believed. The house and pole, for example, are now believed by anthropologist Robin Wright to have been part of a model of Skidegate Village created for display at the exposition that represents the only known example of a North American village to have been "systematically documented by its own nineteenth century residents" (Wright 2015, 381). In addition, a memo in the Chicago Field Museum archive, together with small yellowed labels reading "Hagenbeck" that remain on several of the masks in Vienna, confirm that these were culled from a second Northwest Coast collection made by Adrian Jacobsen and brother Filip between 1884 and 1885 and financed by impresario Carl Hagenbeck. Portions of that collection had been displayed at the fair and purchased by Putnam and Boas in their capacity as exposition representatives (Penaloza-Patzak 2018c, 100–101; Haberland 1989, 185, 190–191).

In 2014 the Hamburg Ethnological Museum—today Museum am Rothenbaum—re-opened its archive, rendering Jacobsen's extensive collection publicly available for the first time in many years. The materials therein include a series of logbooks in which Jacobsen detailed his travels, beginning with the date, weather, and sea conditions of each day before going on to list the objects he bought or traded, the price paid and, in some instances, the name of the individual from whom they had been acquired. In the future, these documents, a combination of personal travel log and account book, may well provide richer insight than initially expected into pieces he collected that later traveled into museums (Penaloza-Patzak 2018c, 101–103). As these case studies indicate, objects have the potential to continue transforming scientific knowledge and museum constellations long after they have settled in collections storage, and we are just beginning to recognize the potential for combined collections and archival research.

Conclusion

The chain of interactions surrounding the 1893/1894 Columbian Exposition ethnological exchanges marked a seminal episode in the complex story of international Americanist anthropology. The exchanges interlaced the interpersonal, intellectual, and institutional interests of Boas, Putnam, and Holmes, all working in the makeshift Midwestern scientific outpost that was the fair, with those of Heger in Vienna, and Bastian and Seler in Berlin, each of whom was deeply embedded within the "thick space" of their respective European metropolitan museums. The transactions appear to have pleased most parties involved, and quickly led to talk of future collaboration between Bastian and Putnam in particular.[13] In the scope of US urban politics, the fair announced Chicagoans' intention to win their city recognition as a metropole in its own right, and the scientific resources amassed to stage the fair provided both material basis and economic impetus for the subsequent construction of the Field Columbian Museum. The museum and its collections legitimized and anchored the city's claim as a new space for knowledge creation, while the ethnographic exchanges contributed to the shape and direction of a transnational exchange infrastructure.

In addition to bringing Chicago into the metropolitan exchange system, the transactions also established points of convergence for the alignment of anthropological interests in Vienna and Berlin. This was indirect at first, by simple virtue of their mutual participation in the fair, but by the end of 1893 Heger was named one of the first Vienna-based members of the BGEAU (Verhandlungen 1899, 5), and in 1908 he, working closely with Seler in Berlin, organized and hosted the 16th International Congress of Americanists at the NHM in Vienna. Throughout the first decade of the twentieth century, cooperations between US- and Berlin-based anthropologists would continue to far outnumber the engagements of either with their Vienna-based colleagues. Nonetheless, there can be no question that Heger's contributions to the exposition secured broader international recognition for Vienna's ethnographic collections and museum practices, and would serve to ally Vienna's anthropological interests with those binding US- and Berlin-based colleagues for many years to come. The apogee of these collective efforts and this tradition of material exchange would be the International School of American Archaeology and Ethnology in Mexico (ISAAE 1911–1914), organized by Boas, directed in its inaugural year by Seler, and sponsored by, among others, both the Austro-Hungarian and Prussian governments, which received small archeological collections from the school in return for their support. The First World War interrupted this tradition of material exchange, but new research suggests that transnational relations between the metropoles persisted well into the interwar period, for example in the form of extensive donations for anthropological research made to institutions in Berlin, Vienna, and Munich between 1920 and 1927 by the Emergency Society for German and Austrian Science and Art, a US-based organization established by Boas (Penaloza-Patzak 2018b; Penaloza-Patzak 2018c, 183–203, 208–214, 245–272; Penaloza-Patzak forthcoming).

Notes

1. Holmes, July 1894 annotated list, Memos /All Depts. Dept. D K.K. Hof Museum #55, Austria, Vienna, July 25, 1894, Chicago Field Museum (FM).
2. Boas to Unknown, December 19, 1893, Chicago Weltaustellung 1893 (CWA), Ethnological Museum Berlin (EMB); and Holmes, July 1894 curatorial memo, Memos /All Depts. Dept. D, July 25, 1894, Prof. H. Virchow #56, University of Berlin, FM; Boas to Bastian, May 8, 1893, A. Jacobsen Vol I IB W, EMB.
3. Unknown to Bastian, March 11, 1893, CWA, EMB; and Langhamer to Putnam, January 10, 1895, Anthropology Department, Folder No. 58, Grassi Museum, Leipzig, AR, FM.
4. Boas to KMV, April 26, 1891, Am V.13; Grünwedel to Boas, May 10, 1891, Am V.13; and Putnam to Seler, May 29, 1891, CWA, all EMB.
5. Seler to Putnam (draft), May 15, 1891, Am V.13, EMB.
6. Boas to Bastian, March 29, 1893, CWA, EMB.
7. Heger to Putnam, March 6, 1893, Archiv für Wissenschaftsgeschichte, Intendanzakten 1893, Z. 105, Natural History Museum Vienna (NHM).
8. Unknown to Bastian, March 11, 1893; and Boas to Seler, March 31, 1893, both CWA, EMB.
9. Boas to Bastian, November 8, 1893; and Grünwedel notes for the Chicago World's Exposition, November 3, 1893, both CWA, EMB.
10. Boas to Bastian, November 8, 1893, CWA, EMB.
11. Holmes, July 1894 annotated list, Memos /All Depts. Dept. D K.K. Hof Museum #55, Austria, Vienna, July 25, 1894, FM.
12. Boas to Bastian, May 8, 1893, A. Jacobsen Vol I IB W; Boas to Unknown, December 19, 1893, CWA, both EMB.
13. See, for example: Bastian to Putnam, May 19, 1894; and Bastian to Putnam received November 24, 1894; both in Memos /All Depts. Dept. D, Prof. A. Bastian #57, Berlin, Germany, July 25, 1894, FM.

Bibliography

Baedeker, K. (1896). *Baedeker's Austria, Including Hungary, Transylvania, Dalmatia, and Bosnia*. Leipzig: Karl Baedeker.

Baedeker, K. (1897). *Northern Germany as Far as Bavarian and Austrian Frontiers*. Leipzig: Karl Baedeker.

Brantz, D., Disko, S., and Wagner-Kyora, G. (eds.) (2012). *Thick Space: Approaches to Metropolitanism*. Bielefeld: Transcript.

Bunzl, M., and Penny, G. (eds.) (2006). *Worldly Provincialism: German Anthropology in the Age of the Empire*. Ann Arbor: University of Michigan Press.

Byrne, S., Clarke, A., Harrison, R., and Torrence, R. (eds.) (2011). *Unpacking the Collection: Networks of Material and Social Agency in the Museum*. New York: Springer-Verlag.

Cole, D. (1985). *Captured Heritage: The Scramble for Northwest Coast Artifacts*. Seattle: University of Washington Press.

Czeike, F. (2004). *Historisches Lexikon Wien* (Vol. 3 Ha-La). Vienna: Kremayr & Scheriau/Orac.

Darwin, Ch. (1886). *The Origin of the Species by Means of Natural Selection, or the Preservation of Favoured Races in the Struggle for Life* (6th ed.). London: John Murray.

Dorsey, G. A. (1899). Notes on the Anthropological Museums of Central Europe. *American Anthropologist* 1:3, 426–474.

Erzherzog Rudolf (ed.) (1886–1902). *Die österreichisch-ungarische Monarchie in Wort und Bild* (Vols. 1–24). Vienna: k. k. Hof- und Staatsdrukerei, Alfred von Hölder.

Feest, C. (1995). The Origins of Professional Anthropology in Vienna. In: Rupp-Eisenreich, B., Stagl, J., and Acham. K. (Eds.), *Kulturwissenschaften im Vielvölkerstaat: zur Geschichte der Ethnologie und verwandter Gebiete in Österreich, ca. 1780 bis 1918*. Vienna: Böhlau-Verlag, 113–131.

Glimpses of the World's Fair: A Selection of Gems of the White City Seen Through a Camera (1893). Chicago: Laird & Lee Publishers.

Gosden, C., Larson, F., and Petch, A. (2007). *Knowing Things: Exploring the Collections at the Pitt Rivers Museum 1884–1945*. Oxford: Oxford University Press.

Haberland, W. (1989). Remarks on the 'Jacobsen Collections' from the Northwest Coast. In: Peset, J. L. (ed.), *Culturas de la Costa Noroeste de América*. Madrid: Turner.

Heinrich, A. (1995/1996). Von Museum der Anthropologischen Gesellschaft in Wien zur Prähistorischen Sammlung im k. k. Naturhistorischen Hofmuseum (1870-1876-1889-1895). *Mitteilungen der Anthropologischen Gesellschaft in Wien* 125/126: 11–42.

Heinrich, A. (2006/2007). Franz Hegers Reisen und Ausgrabungen im Kaukasus und die Entstehung der 'Sammlung Kaukasischer Alterthümer' in Naturhistorischen Museum in Wien. *Mitteilungen der Anthropologischen Gesellschaft in Wien* 136/137: 107–143.

Hinsley, C. M. (2016a). Anthropology as Education and Entertainment. Frederic Ward Putnam at the World's Fair. In: Hinsley, C. M., and Wilcox, D. R. (eds.), *Coming of Age in Chicago: The 1893 World's Fair and the Coalescence of American Anthropology*. Lincoln: University of Nebraska Press, 1–77.

Hinsley, C. M. (2016b). Ambiguous Legacy: Daniel Garrison Brinton at the International Congress of Anthropology. In: Hinsley, C. M., and Wilcox, D. R. (Eds.), *Coming of Age in Chicago: The 1893 World's Fair and the Coalescence of American Anthropology*. Lincoln: University of Nebraska Press, 99–109.

Hinsley, C. M., and Wilcox, D. R. (eds.) (2016). *Coming of Age in Chicago: The 1893 World's Fair and the Coalescence of American Anthropology*. Lincoln: University of Nebraska Press.

Hirschberg, E. (1903). *Statistisches Jahrbuch der Stadt Berlin* (Vol. 27). Berlin: P. Stankiewicz' Buchdrukerei.

Holovatch, Y., Kenna, R., and Turner, S. (2017). Complex Systems: Physics Beyond Physics. *European Journal of Physics* 38, 1–22.

Jacknis, I. (1985). Franz Boas and Exhibits: On the Limitations of the Museum Method in Anthropology. In: Stocking, G. (ed.), *Objects and Others: Essays on Museums and Material Culture*. Madison: University of Wisconsin Press, 75–111.

Johnson, R. (ed.) (1898). *A History of the World's Columbian Exposition in Chicago in 1893* (4 vols.). New York: D. Appelton and Company.

Krieger, K. (1973). Abteilung Afrika. In: Krieger, K. and Koch, G. (eds.), *Ethnologisches Museum Berlin. Hundert Jahre Museum für Völkerkunde Berlin* (Baessler-Archiv, Neue Folge, Vol. 21). Berlin: D. Reimer, 101–140.

Nichols, C. A. (2018). The Smithsonian Institution's 'Greatest Treasures': Valuing Museum Objects in the Specimen Exchange Industry. *Museum Anthropology* 41:1: 13–29.

Penaloza-Patzak, B. (2018a). An Emissary from Berlin: Franz Boas and the Smithsonian Institution, 1887. *Museum Anthropology* 41:1, 30–45.

Penaloza-Patzak, B. (2018b). Das Emergency Society for German and Austrian Science and Art, 1920–1927. Eine Einführung in eine beinahe unbekannte Hilfsorganisation und der Mehrwert ihrer Erforschung. In: Feichtinger, J., Klemun, M., Surman, J., and Svatek, P. (Eds.), *Wandlungen und Brüche. Wissenschaftsgeschichte als politische Geschichte*. Göttingen: V & R unipress, 125–132.

Penaloza-Patzak, B. (2018c). Guiding the Diffusion of Knowledge: The Transatlantic Mobilization of People and Things in the Development of US Anthropology, 1883–1933. PhD dissertation, University of Vienna.

Penaloza-Patzak, B. (forthcoming). Friends in Deed: Allies in the Interwar Struggle for 'German' Science and Art. In: *Academies and World War I: The Aftermath* [special issue] (Acta Historical Leopoldina).

Penny, G. H. (2002). *Objects of Culture: Ethnology and Ethnographic Museums in Imperial Germany*. Chapel Hill: The University of North Carolina Press.

Pierce, B. L. (1957). *A History of Chicago: The Rise of a Modern City 1871–1893* (Vol. 3). Chicago: University of Chicago Press.

Rand, McNally & Co.'s A Week at the Fair: Illustrating the Exhibits and Wonders of the World's Columbian Exposition with Special Descriptive Articles. (1893). Chicago: Rand, McNally & Co. Publishers.

Ranzmaier, I. (2011). The Anthropological Society in Vienna and the Academic Establishment of Anthropology in Austria, 1870–1930. *History of Anthropology Annual* 7, 1–22.

Ranzmaier, I. (2013). *Die Anthropologische Gesellschaft in Wien und die akademische Etablierung anthropologischer Disziplinen an der Universität Wien, 1870–1930*. Vienna: Böhlau-Verlag.

Report of the Massachusetts Board of World's Fair Managers. (1894). Boston: Wright & Potter Printing Co.

Sedlaczek, S., Löwn, W., and Hecke, W. (1902). *Statistisches Jahrbuch der Stadt Wien in das Jahr 1900* (Vol. 18). Vienna: Verlag des Wiener Magistrats.

te Heesen, A., and Vöhringer, M. (eds.) (2014). *Wissenschaft im Museum—Ausstellung im Labor*. Berlin: Kadmos Verlag.

Verhandlungen der Berliner Gesellschaft für Anthropologie, Ethnologie und Urgeschichte. *Zeitschrift für Ethnologie* 31 (1899).

Westphal-Hellbusch, S. (1973). Zur Geschichte des Museums. In: Krieger, K., and Gerd Koch, G. (eds.), *100 Jahre Museum für Völkerkunde Berlin* (Baessler-Archiv Beiträge zur Völkerkunde, Vol. 21). Berlin: D. Reimer, 1–100.

Wilcox, D. R. (2016). Anthropology in a Changing America. In: Hinsley, C. M., and Wilcox, D. R. (eds.), *Coming of Age in Chicago: The 1893 World's Fair and the Coalescence of American Anthropology*. Lincoln: University of Nebraska Press, 125–232.

Wright, R. K. (2015). Skidegate Haida House Models. In: Kan, S. (ed.), *Sharing Our Knowledge: The Tlingit and Their Costal Neighbors*. Lincoln: University of Nebraska Press, 380–393.

Part Four
Focus on Vienna 2
Sciences and Publics

9 Talking About Popular Science in the Metropolis

Learned Societies, Multiple Publics, and Spatial Practices in Vienna (1840–1900)

Johannes Mattes

Introduction

Modern cultures of urban living cannot be separated from the sciences and technology. Processes of urbanization and spatial centralization went hand in hand with the institutionalization of scientific societies, universities, libraries, or museums, but new ways of scientific administration/governance and specific practices of communication also developed that interacted between different layers of academic science and the public sphere. All metropolises connect people, cultures, and practices of knowledge. As multiple spaces, major urban centers represent intersections of various social, geographical, architectural, and epistemological concepts of space. Representing dynamically changing meeting-places, metropolises and their specific mediality and polyvalent imagery connected different communities.

According to the relational spatial theory of Henri Lefebvre (1992), which describes the constitution of space as a result of a permanent rearrangement and collocation of bodies and objects, metropolises are also constituted by their transitional position and the dynamics of the global and the local, the dependency of centers and peripheries, and their relations with other cities. Postcolonial studies (Ashcroft et al. 2000, 113–114) have also criticized the dichotomy of metropolis and marginality, pointing out that the dualism of the center-periphery concept denies the existence of local knowledge production and regional scientific centralism (see also Hochadel, Chapter 3 in this volume). Therefore, major urban centers of knowledge cannot be seen apart from their respective peripheries, but also have to be scaled and examined according to their location, size, function, centrality, consolidation, relation to other cities as well as their inherent material, social, and symbolic dimensions of knowledge.

Since the 2000s, historians of science have assigned cultural brokers, interpreters, and translators a more significant role in global and urban knowledge production and transformation. In his book *Science in the Public Sphere*, Agustí Nieto-Galán (2016, 18) pays special attention to the intermediate role of science popularizers and argues that "the struggle between professional science, associated with experts and orthodoxy, and amateur science, associated with laypeople and heterodoxy, was, and continues to be, aired in the public sphere at different levels in a constant struggle for cultural hegemony." For the French-Indian scholar Kapil Raj (2016), go-betweens and intermediators hold this whole complex world together by carrying and brokering spatialized knowledge between different communities. This concept could also be transferred to the metropolitan fabric. In this context, new transportation infrastructures in metropolises were not the only spaces in-between. Building bridges and circulating between mutable fields and cultures of knowledge, go-betweens, mediators, and also physical objects served as cultural translators. They moved in spaces of uncertain or changing boundaries, constituted or served as resources for networks, and thus helped to form and cement the urban fabric. According to Bernhard Dotzler and Henning Schmidgen, such "spaces in-between" represent "interfaces, intervals and gaps, where elementary processes of knowledge production take place", where discourse and communication meet material culture and knowledge is translated from one culture to another (Dotzler and Schmidgen 2008, 7).

For Ulrike Felt the interrelation between science and the public is not limited to knowledge transmission. Both spheres are constituted through ongoing processes of demarcation between scientific disciplines, science and non-science, leading and supporting "actors", science and politics: "Space means demarcation or differentiation of other places, refers to specific content and structural features, symbolizes the existence of 'access requirements' or prefers certain actor constellations to others. Thus, these places of encounter and confrontation with science also constitute new forms of publicity" (Felt 2000, 202–203).

In speleology, the scientific study of caves—a metadisciplinary field of research, where scientific and popular knowledge of geology, geography, zoology, paleontology, botany, prehistory, and cultural anthropology interlinked—transitional zones of knowledge and the circulation of go-betweens in metropolises acquired particular significance. Mainly practiced by travelers, scholars, and amateurs in the mountainous peripheries of European major cities, the exploration and scientific examination of caves as archives for natural and human history developed as a primarily urban phenomenon, enhanced by popular books such as Jules Verne's *Journey to the Center of the Earth* (1864) or Georg Hartwig's *The Subterranean World* (1871). Of course caves were also often visited by locals, but the idea of recognizing natural cavities as places of knowledge mainly emerged in major cities, accompanied by the awareness that metropolises consist of both above-ground infrastructures and their subterranean counterparts like subways, passageways, transportation and communication hubs, and sewer systems (Pike 2005; Barták et al. 2007). In contrast to horizontal movements like expansion, migration, circulation, and diffusion, metropolises grow into the sky and also into the subterranean realm. As described by Mattes (2015, 210–217), the emergence of these new subterranean spaces in the underground imagination of late nineteenth-century metropolises went hand in hand with the scientific examination of caves and their exploitation for touristic purposes. Especially in Vienna and Paris, where the first influential societies for cave study were founded in the second half of the nineteenth century and speleology was subsequently institutionalized in the form of state-owned research institutes, cultural and scientific discussions of caves became sites of interaction for different scientific disciplines that required the cooperation of professional scientists and amateurs to accomplish. It is evident that cave study represents neither an influential nor important field of research, but studying its epistemic setting in Vienna is important because it is closely related to the development of Vienna as a metropolis of scientific impact.

Following Juraj Kittler (2013, 285), who argues that a major city generates specific formats of communication and the "urban form and the semiotic codes embedded in its architecture turn the city itself into an ultimate medium", this chapter examines (1) the various modes of communication with multiple publics and fields of knowledge as a signature of nineteenth-century's metropolises, (2) the function of popular scientific societies, urban go-betweens, and their spatial settings for the development of knowledge-based public spheres, and (3) the role of cave study as an example of a field of research that emerged at the margins of different academic disciplines and the public sphere. Concentrating on two selected representatives—Adolf Schmidl and Franz Kraus—and their learned communities, the chapter further examines the function of spaces in-between for the communicative setting in metropolises and its impact on the accumulation of knowledge in urban environments.

1. Metropolises and Science Communication

In the second meeting of the newly founded Imperial Academy of Sciences in Vienna (1847), its members discussed practical questions of science communication.[1] Due to his incautious handling of information, the geologist Wilhelm von Haidinger (1795–1871) was strongly attacked by his Viennese colleagues, among them the Orientalist and Academy president Joseph von Hammer-Purgstall and the chemist Anton Schrötter von Kristelli. Because Haidinger had taken the liberty of reporting on the first scientific lectures given at the academy in a newspaper (Anonymous 1847a), he was imputed to have harmed the "fundamental conditions for all scientific activity inside the academy" (Imperial Academy of Sciences 1847). In addition, Haidinger had also reported on the academy's lectures in a non-state-authorized private society called *Freunde der Naturwissenschaften* ("Friends of the Natural Sciences"), an informal club with a quite popular scientific lecture series that he had established in 1845 along the lines of similar societies in Edinburgh, where he had spent two years as a young scholar (Anonymous 1847b). In his apology, Haidinger argued: "It would be an outrageous tyranny, if the academy seeks to prevent its own members from making known what every unnamed, unknown correspondent in the general newspaper is allowed to do. . . . I would like to be a limb of a living body. . . . The academy was founded for the enlargement and dissemination of scientific knowledge, it is a means, not an end in itself" (Imperial Academy of Sciences 1847). In the following close vote, the members decided to publish regularly short reports on the academy's scientific activity in the newspaper *Wiener Zeitung*, an official organ of the state.

Communicating science to the multiple publics that developed during a period of political transformation remained a rather problematic issue for state-founded learned societies in Vienna even after 1848, when censorship was mainly lifted in the Habsburg Empire. Frequently, reform of press censorship and scientific societies were discussed together (Anonymous 1848). In 1869, the geologist and Academy member Ami Boué published a 78-page polemic entitled, "A free word about the Imperial Academy of Sciences" (Boué 1869, 10–11, 23), where he sharply criticized his own learned society:

"The weekly lecture series at the academy, the first scientific body in the state, usually have no attendance. . . . Instead, all other scientific associations in Vienna are very much in vogue. . . . It must be said that most societies do not have large conference rooms and could not cover such expenses. We would be in a great deal of embarrassment, if—instead of the usually empty auditorium—many listeners frightened us. . . . In the current practical age, where everything is public, where everything is discussed in the newspapers and the state expenses of an institution are critically illuminated from all sides, it is absolutely necessary to add instructive sessions to our lecture series. . . . Some scholars tend to give long lectures, but for discussions they fear a violation of 'academic decorum'! Debates would allow non-members or newspapers to attack them. But such controversies have often existed and were of highest interest for larger audiences."

Boué's certainly exaggerated remarks were published in the context of a long-lasting discussion about a reform of the Academy's bureaucratic body. Around 1868, the reform debate reached a new peak and was pushed forward by a group of scholars related to the Geological Survey, who demanded an opening of the Academy to the public similar to other scientific societies. In this respect, the title of Boué's paper "A free word . . ." acquired a second meaning, namely "free communication". Similarly, it refers to the growing relevance of popular science, publicity, and non-professionals in research.

These quotations about scholarly communication and the dissemination of knowledge suggest, first of all, that scientific communication was only partially plannable or controllable in this period. Especially in major cities with a high concentration of scientific institutions and media, communication also happened outside the disciplining formats of conferences, lectures, scientific journals, or letters. In public spaces and media, knowledge could circulate anonymously and without a clear attribution to its originator. Second, the scientific reputation of a learned society did not necessarily correspond to a large audience or free access to knowledge. Likewise, a small audience was not necessarily an indication of scientific excellence. Frequently, a scholar gave the same lecture to different societies and the content was "translated" into different contexts, which accelerated the concentration and circulation of knowledge.

Third, it is not possible to identify the players at this time as representing either "science" or a "public" that could be regarded as homogenous in any sense. Similarly, professionalism and amateurishness in nineteenth century science represent quite mutable entities and the fact that in the Habsburg Empire many scholars were employed by the state may have delayed processes of science professionalization (Coen 2018). In metropolises, multiple publics and communities of discourse between various spheres of science and the public, critiquing, questioning, and negotiating matters of scientific meaning correspond to different formats and modes of communication. Fourth, while some scholars mobilized public interest in scientific research as a resource to acquire support for their own projects—in particular in fields of research, where the observations of numerous non-professionals were needed to accomplish research projects—others regarded this incautious handling of scientific knowledge as a danger for their own research.

It was at this very time that the first professionals in science popularization emerged. According to Andreas W. Daum (1998), who has examined this process in the case of bourgeois culture in Germany, many of these professionals worked as scientific journalists in major European cities. Nevertheless, in the Habsburg Empire of the early nineteenth century, strict censorship carried out by an often-overlapping conglomerate of functionaries, censors, and scholars had impeded the development of independent journalism and scientific research during the *Vormärz* period (1815–1848). According to Marianne Klemun (2020), scientific reports in newspapers were often written anonymously by scientists themselves (Anonymous 1849). As Adolf Schmidl reported in 1852: "Even though printing in Vienna is now less than in any city of the same size . . . , the reason for this lies chiefly in the as yet unresolved real shyness of the Austrian from any publicity, and the motto '*novem in annos*' was observed to the limit here. Therefore, literature in Vienna has not been degraded to such formal forms of employment as elsewhere" (Schmidl 1852, 33). Schmidl, a geographer, was one of these early Viennese representatives of science popularization, who became editor of the journal *Österreichische Blätter für Literatur und Kunst, Geschichte, Geographie, Statistik und*

Naturkunde (1844–1848), a popular learned periodical in the Habsburg Empire. Although Schmidl had published unauthorized articles (Anonymous 1847c) about the foundation of the Imperial Academy of Sciences in his journal, the Academy's members did not hesitate to elect him to the position of actuary (head of administration). From their viewpoint, it seemed easier to control Schmidl's urge to publish, if he were employed by the Academy.

Fifth, scholarly communication is not only determined by social and epistemic, but also by spatial and architectural conditions. The content of lectures or newspapers becomes rather ephemeral, when compared to the lasting power that both scientific infrastructure, intermediating technologies, and the urban form hold over learned communication. For Habermas (1989), the physical public space and the mediated public sphere can be regarded as two sides of the same coin. The urban form of the modern metropolis with its assemblies, lecture halls, club houses, museums, collections, and coffeehouses has to be considered as the context within which popular science, new modes of participation and scientific communication developed. Accordingly, despite the relative lack of construction activity in the *Vormärz* period, the meetings of the first scientific societies in Vienna did not exclusively take place as private circles in the homes of particular scholars, pubs, or in the so-called *Bäckerviertel* ("Baker's Quarter") next to the university, but also occurred in the administrative buildings newly erected between the "Glacis" and the present-day inner districts. These included the Mint (*Haupt-Münzamt*), the Polytechnic Institute, and the Vienna Regional Court. Thus, such places had expanded the range of potential sites of knowledge acquisition and exchange already in the first part of the nineteenth century (Csendes 2003). This process was further accelerated during the 1860s, when the demolition of the traffic-impeding city fortifications and the construction of the Ringstrasse boulevard went hand in hand with bourgeois demands for a right of assembly, which would simplify the founding of associations.

2. Shaping the Metropolis: Scientific Societies and the Public Space

Intermediating between science and the public sphere, learned societies took over several tasks: First, they created an internal communication platform for their members through meetings and printed circulars. They took control of free spaces in-between, both in communication and in the urban infrastructure, and addressed goals that were not yet pursued by other research institutes or traditional institutions such as the state or the church. Second, they also channeled communication with the outside world, mediated between regional and cross-national knowledge cultures and generalized local knowledge. In doing so, associations played a key role in the organization of the developing civil society during the reduced influence of a corporative state structure; they dealt with the individual interests of their members and formed a social group identity. The distinguishable from informal associations by their continuity, publicity, statutory framework, and voluntariness, these early societies dedicated themselves to the acquisition of knowledge, social welfare, and different kinds of culture; by these means the emerging urban bourgeoisie was able to take over these fields and develop its own cultural identity (Klemun 1998, 13).

This means that, in the Habsburg Empire, science became a bourgeois practice not earlier than in the first half of the nineteenth century and was transformed step by step and with multiple speeds into a public good. This quite dynamic process, negotiated at the intersection between science, politics, and religion, is closely related to the increased impact of the public sphere and the emergence of learned societies that served as a catalyzer of this development. While the foundation of patriotic, commercial, and agricultural societies reached its peak in Germany in the second half of the eighteenth century, reading associations had a peak around 1800 (Daum 1998, 85–88). In the Habsburg Empire, similar processes happened about 30 years later and with a higher degree of spatial differentiation. While in the German territories 12 associations (excluding Academies of Sciences) developed in the second half of the eighteenth century, Vienna hosted only one association, the Cosmographic Society, that existed between 1790 and 1797 and was subsequently transformed into a state-related cartographic institute (Anonymous 1810). Although the suppression of associations played an important role during the *Vormärz* period, the state administration was not generally negative about the need for increased scholarly exchange and supported associations for the promotion of the sciences and arts, agriculture, industrialism, and other public interests (Stubenrauch 1857). The mutual interactions of state, civil society, and scientific interests in the monarchy led to different foundations or respectively founding attempts, due either to governmental or to private (state-authorized) initiatives (Kadletz-Schöffel 1992). Thus, the focus on patriotic or economic projects such as agriculture, pomology, hydraulic engineering, mining, geography, cartography, geognosy, or medicine, and the close connection to the sovereign accelerated the formation of scientific disciplines, which became institutionalized around 1850 through the founding of the Imperial Geological Survey or the Central Meteorological Institute.

In this regard, early scientific societies in Vienna could be understood as spaces in-between, where state and bourgeois interests met—often not without conflict. These included the Society for Physicians (*k. k. Gesellschaft der Ärzte*) and the Horticultural Society (*k. k. Gartenbaugesellschaft*), both founded in 1837, the "Reading Club for Politics and Jurisprudence" (*Juristisch-politischer Leseverein*), founded in 1841, and the "Friends of Natural Sciences" (*Freunde der Naturwissenschaften*, 1845–1850), mentioned previously, which only existed for five years, but whose members came together again in the newly founded Imperial Geological Survey (*Geologische Reichsanstalt*) in 1849 (Klemun 2020).

Originally an unofficial association that had developed from regular meetings led by Haidinger in the Mining Museum (*Montanistisches Museum*), the "Friends of Natural Sciences" understood themselves as a liberal alternative to the autocratic and state-initiated Academy of Sciences and became a hub for the foundation of other societies for natural science in the Habsburg Empire (Haidinger 1869). In contrast to the Academy, subscribers financed the printed proceedings of the "Friends of Natural Sciences" and state officials never authorized the statues of the society. As one report stated:

"In Vienna, we have several associations, which deal with practical branches of natural sciences. However, there was no scientific society that dealt with all branches of natural sciences, especially with the extension of science, and was accessible to everyone. While these clubs develop a more practical tendency to move within the sphere of different disciplines and have no general accessibility, they cannot satisfy the ideas of a society for general dissemination and expansion of the sciences in all their directions, which we have grasped and expressed. . . . An academy must necessarily be confined to a number of members, but the number of devotees of the natural sciences is, as happens here in our lively interest, a very important one, and ideally an unlimited one. Our union [, the "Friends of Natural Sciences"], on the other hand, has a completely different purpose and auditorium. It relies on its members on such individual works, which correspond to the detailed study of each individual. It gives younger researchers in particular the opportunity to share the results of their investigations. . . . It will contribute to the general dissemination of love for the natural sciences to that exalted study, which elevates us and inspires us, for the good and the beautiful, in a manner that gives everybody unlimited access" (Hammerschmidt 1848, 276–277).

"Access" was also a key word in Adolf Schmidl's well-known guidebook, "Vienna and its environs with particular regard to its scientific institutes and collections" (1852)—an important source for the transformation of the city at the turn to the second half of the nineteenth century. As manuals for understanding urban space, guidebooks played a vital role in the inscription of the metropolitan fabric. As narrative instruments, they were not simply tools to facilitate orientation or transportation. In their representative function, they made concessions to their sponsors, shaped the public image of science in the metropolis, and inscribed scholarly places of remembrance in the form of featured landmarks and street names. Reprinted in seven German and two French editions between 1837 and 1858, Schmidl's guidebook experienced a profound modification, which the author explained by the transformation of the city's visible memorabilia of science during these two decades. For each edition, he wrote, "it is my duty to travel round and round, to visit all public institutes and record the scale of current deviation. . . . Austria and in particular the imperial city of Vienna have entered a period of continuous redesign and each description is only a section of Vienna's intermediate state" (Schmidl 1852, VI–VII). In this context, transportation routes acquired new meanings as meeting places and transitional zones for the city's population. Accordingly, Schmidl (1852, 17) became an advocate for the reconstruction of the space between the suburbs and the city's fortifications, to create a "public space with theaters, a stock exchange and more than 30 representative buildings". Defining urban mobility and the transformation of knowledge by spatial resetting as an immanent factor of major cities, he provided in his guidebooks not only a modern city map, but also a distinct table of urban research institutes, scientific collections, and their accessibility (opening hours, contact person, registration).

The foundation of new learned societies was stimulated after the revolution of 1848, when censorship was abolished and new laws governing associations were promulgated in 1854 and 1867. This enabled the foundation of new scientific societies such as the Zoological-Botanical Society (*k. k. Zoologisch-Botanische Gesellschaft*), founded in 1851; the Geographical Society (*k. k. Geographische Gesellschaft*), founded in 1856; the Association for the Dissemination of Scientific Knowledge (*Verein zur Verbreitung wissenschaftlicher Kenntnisse*), founded in 1860; the Anthropological Society (*Anthropologische Gesellschaft*), founded in 1870; and the Scientific Club (*Wissenschaftlicher Club*), founded in 1876 (Hye 1992; Raptis 1998).

Table 9.1 provides a chronological overview of the scientific associations in Vienna and the social status of their initiators. In the *Vormärz* period, the foundation of associations was closely linked to public authorities and intended the promotion of specific "useful" research fields for the good of the state. The associations that were founded between 1839 and 1848 represent an exception. They were supported by the educated and commercial middle class, pushed liberal ideas, and became the starting points of the revolution in Vienna. After 1848, the bourgeoisie and their scientific societies sought again the proximity to the state. However, several new foundations go back to societies in the *Vormärz* period like the Friends of the Natural Sciences and represent to some extent successor organizations. k. k. Associations such as the Geographical Society made the imperial motto '*viribus unitis*' their own and even printed it on the covers of their journals.[2] They received generous support from the authorities, high aristocracy, and the imperial family, and saw themselves as state, cross-regional, and transnational research organizations. It was not until 1870 that societies supported by students and university staff emerged. Due to an increasing strength of the commercial middle class, an interest in the popularization of sciences (since the 1860s) and national politics, associations were founded that were less bound to state politics.

Table 9.1 Scientific Societies in Vienna, 1790–1913, listed by founding date and classified according to the social status of their initiators.

Founding Date	Association Name	Founders' Status*
1790	Cosmographical Society	5
1807	k. k. Agricultural Society in Vienna	1
1837	k. k. Horticultural Society	2
1837	k. k. Society of Physicians	2
1839	Commerce Club of Lower Austria	5
1841	Reading Club for Politics and Jurisprudence	4
1845	Botanical Exchange Club	5
1845	Friends of Natural Sciences	4
1848	Austrian Engineers Club	5
1851	k. k. Zoological-Botanical Society	2
1853	Antiquity Club in Vienna	3
1856	k. k. Geographical Society	2
1860	Association for the Dissemination of Scientific Knowledge	3
1862/1873	(German and) Austrian Alpine Club	3
1864	Club for Regional Studies of Lower Austria	3
1865	Austrian Society for Meteorology	3
1868	Military-Scientific Club	3
1869	Chemical-Physical Society	6
1869	Austrian Tourist Club	3
1870	Numismatic Society	3
1870	Anthropological Society	3
1872	Academic Club of Natural Historians (Scientific Club at the University)	6
1875	Club of Geographers at the University of Vienna	6
1876	Scientific Club	5
1879	Club for Cave Study	3
1883	Vienna Electro-Technical Society	5
1894	Society of European Ethnology	3
1895	Society for the Promotion of Natural Research of the Orient	3
1903	Club for the Promotion of Natural Research of the Adriatic	3
1907	Geological Society in Vienna	6
1913	Vienna Prehistoric Society	6

Source: Compiled from the information in the first issues of the societies' journals.

*1 = State; 2 = State-privileged associations, founded by members of the bourgeoisie, state officials, and aristocrats; 3 = Associations founded by members of the bourgeoisie, state officials, entrepreneurs, and mostly supported by so-called protectors (members of the imperial family); 4 = Associations without a distinct relation to the state, founded by various members of the bourgeoisie; 5 = Associations whose initiators had a technical or commercial background; 6 = Associations founded by university members, researchers, or students.

By 1890, there were already 355 scientific associations with 45 sections in the different provincial capital cities of the Cisleithanian crown lands of the empire (k. k. Statistische Central-Commission 1892). Seventy-seven learned societies were situated in Vienna alone, followed by Prague (45), Graz (13), Chernivtsi (then called Czernowitz) (10) and Lviv (then called Lemberg) (9), Brno (then called Brünn) (9), Innsbruck (9), Ljubljana (7), and Triest (7). Surprisingly, around 150 scientific societies were not located in Vienna or the monarchy's regional capitals, but in minor towns (mostly in Bohemia, Galicia, and Moravia). Aside from secondary schools (Gymnasien) and local museums, they often represented the only learned institutions in these smaller cities and stood in close contact with related associations in regional capitals or Vienna, with which they established an active exchange of literature and scientific objects (Klemun 2003).

As communicative circles of exchange, learned associations not only played a crucial role for the spatial dissemination of knowledge between various fields of research, but were also involved in the production and concentration of new knowledge through its translation from one context to another. With the help of these associations, Viennese bourgeois society was able to create a dense network of personal connections that shaped everyday life. Frequently, a scholar (either professional or amateur) belonged to several such societies, where he attended lectures up to once a week (Hye 1988). Therefore, it is no wonder that the number of members of these scientific societies increased quite dramatically. Ten years after their foundation, the Zoological-Botanical Society already had 1,010 and the Geographical Society 419 members (Frauenfeld 1863; Fötterle 1868). However, annual membership fees—for example 4 florin (1858) for the Zoological-Botanical Society—were high, so membership in scientific societies was also a question of financial resources. Therefore, many members, once they had paid the entrance fees, did not maintain their memberships continuously. To put it in a nutshell: To be part of what the scientific culture of Vienna understood as "the public" in the second half of the nineteenth century, financial well-being was a prerequisite. Large parts of society, especially the social underclass, were excluded from this new concept of public space and thus also from the culture of scientific communication.

Despite the close relationships among these learned societies, their particular functions for scientific communities were quite diverse:

1. Some associations, like the Anthropological Society, pursued the establishment of the field as an academic discipline and the foundation of a chair at the University of Vienna (Ranzmaier 2013; Pusman 2008).
2. Others, such as the Geographical and the Zoological-Botanical Society, which required the observations of numerous non-professionals to accomplish research projects, tried to establish platforms of

exchange between professionals and amateurs. By attributing reputation to their members, these societies represented a pool of volunteers as well as an audience for established scholars, who engaged members, mainly interested state officials, as promotors for the realization of their research projects. For example, on the occasion of the fiftieth anniversary of the Geographical Society, the festival lecturers mentioned the main function of the association (Tietze 1907, 80):

> "Haidinger [as its first president] saw the purpose of the society as creating a voluntary point of union for those, who wanted to promote the interests of geography and participate in the progress of geographical knowledge. . . . So, it was not just intended to be an association of professional scholars. . . . To understand how modern this was at the time, it must be remembered that Vienna had a big gap in its intellectual life and that the great number of associations and institutions that are nowadays involved in the popularization of knowledge did not exist."

In contrast to the exclusive form of scholarly organization in the Imperial Academy of Sciences, these societies emphasized their inclusiveness. Similarly, Georg von Frauenfeld (1851, 2–3) described the purpose of the Zoological-Botanical Society during its founding meeting: "Only cooperation doubles, multiplies the power. . . . Cooperation should be realized through lively fellowship, confidential communication about the development of research, encouraging discussion of intended works and continuous renewal of mutual understanding among the members. Everybody should be included. The whole [community] relies on the work of individuals."

3. Interdisciplinary associations like the Club for the Dissemination of Scientific Knowledge or the Scientific Club can be regarded as responses to the foundation of disciplinary learned societies and as locations for efforts to popularize scientific knowledge beyond such organizations. Their representatives "pointed out that although the disciplines possess their associations, which allow the public access to their libraries, no center has yet been created in which the exchange of ideas between the scholars and other educated circles is possible" (Kanzlei des Wissenschaftlichen Club 1876, 1).

4. In addition, specialized societies such as the Club for Cave Study (1879) developed in spaces between already established fields of knowledge, endeavoring to coordinate research projects and promote public engagement with science (Mattes 2019). Their representatives, like Richard Issler, editor of the newspaper *Neue Deutsche Alpenzeitung*, or Josef Szombathy, assistant curator at the newly founded Natural History Museum in Vienna, came from different fields of science and the public media.

Transformations of political practice after the revolution of 1848 also necessitated a reconstruction of urban space and its transitional zones. As Schmidl (1852, 9) noted in his guidebook: "It shall not be denied that in contrast to other major cities Vienna has not enough representative buildings. Now, this has become a severe problem, especially because the publicity and orality of each procedure and the possibility to found societies necessitate more public space and meeting places."

The space requirements for these early scientific societies in Vienna were quite high. Some of them, like the Political-Juridical Reading Club, held libraries of up to 10,000 volumes (Brauneder 1992). Others created enormous archives and scientific collections, organized research trips and expeditions to the Balkans and overseas, published yearbooks or regular scientific journals, and hosted popular lecture series with audiences that exceeded even those of university lectures. In Vienna at the dawn of the second half of the nineteenth century, the shortage of public accessible urban space, in particular for large audiences, led to the use of privately owned public open space—a phenomenon that became significant for spatial practices of science in Vienna. The societies often sought spatial proximity to established political, commercial, and scientific institutions. Some associations used the facilities of museums and state-owned research institutions, and particularly the Parliament of Lower Austria provided free of charge shared space for several scientific societies in its Palais in the Herrengasse, where the revolution of 1848 had broken out. Even the Imperial Academy of Sciences made its meeting facilities regularly available for the Geographical Society, after the Academy was awarded the former university building in 1857. This multiple usage of space was significant for the communication and exchange between these scholarly associations, especially in the second half of the century. After the opening of the Palais Eschenbach (Eschenbachgasse 9–11, close to the Ringstrasse) in 1872, which became the meeting place of many learned societies, the shortage of space was relieved. In 1876, for example, the Scientific Club, initiated by the geologist Ferdinand von Hochstetter, rented not less than six assembly rooms in the Palais Eschenbach, which became venues for weekly lectures on different aspects of (popular) science (Kanzlei des Wissenschaftlichen Club 1876, 5–6). Since the Scientific Club did not charge other learned associations for the use of its boardrooms according to its statutes, the club became an important intersection of the societies' activities in Vienna.

Figure 9.1 provides an overview of important meeting rooms used by scientific associations and their spatial distribution in the city. During the *Vormärz* period, the seat of the Estates of Lower Austria (No. 1) served as the main meeting place. Since 1850, the facilities of scientific institutions (No. 2–5), located outside the inner city, were additionally used. In case of larger events, cultural associations such as the Vienna Music Club (*Musikverein*) (No. 6) even made their auditoria available. Since

Figure 9.1 Locations of important meeting rooms used by scientific societies between 1848 and 1914. The dashed line shows the course of the Ringstrasse. Map basis: Blumauer, Wien 1875, Franzisco-josephinische Landesaufnahme (Blatt 4757–1-a).

Source: Österreichisch-Ungarische Monarchie, Militärgeographisches Institut/Public domain via Wikimedia Commons: https://commons.wikimedia.org/wiki/File:Aufnahmeblatt_4757-1-a_1875_Wien_Innenstadt.jpg. Download date: 28 March 2020.

Notes: 1. Seat of the Estates of Lower Austria; 2. Mining Museum; 3. Geological Survey; 4. Botanical Museum of the University; 5. Polytechnic Institute (from 1872: Higher Technical Institute); 6. Seat of the Vienna Music Club (*Musikverein*); 7. Academy of Sciences; 8. Former main building of the University of Vienna; 9. Seat of the Horticultural Society; 10. Academic Gymnasium; 11. Academy of Commerce; 12. Palais Eschenbach, seat of the Austrian Engineers Club and the Commerce Club of Lower Austria; 13. Vienna Stock Exchange; 14. New main building of the University of Vienna; 15. Natural History Museum; 16. Art History Museum; 17. Seat of the Society of Physicians in Vienna; 18. Seat of the Military-Scientific Club; 19. Seat of the 'Urania', an association for adult education.

the 1860s, the former University Square (*Universitätsplatz*), enclosed by the Academy and old University buildings (No. 7–8), represented the main meeting point. Until the turn of the twentieth century, most associations relocated their assemblies and lecture series to the new facilities in the representative buildings of scientific and commercial institutions (No. 9–19) that were constructed along the Ringstrasse between 1870 and 1910.

3. The Role of Go-Betweens and Entangled Fields of Knowledge

In the following sections of this chapter, I wish to show in greater detail how these intermediators played a key role in the spatial concentration of knowledge in the metropolitan environment of Vienna, focusing on the activities of Adolf Schmidl and Franz Kraus. In the case of nineteenth-century Vienna, public space developed from previous spaces in-between, described above, where new forms of communication appeared and negotiation processes between science and the public sphere took place. While public space was dominated until the end of the *Vormärz* period mainly by representative issues of the political elite, new forms of publicity arose after 1848. However, it must be stated clearly that these new scientific societies entangling diverse fields of knowledge did not represent a public space in today's meaning. Their printed proceedings, journals, and lectures often addressed different levels of readership and audience—both society members (semi-public) and a broader audience (colleagues or the learned middle/upper class). In particular, extraordinary (public) lectures at the University of Vienna and journals like Schmidl's *Österreichische Blätter* (Schmidl 1847) became initial points for this new understanding of publicity. Bringing together "the intellectual power of the monarchy [such as Christian Doppler, Franz Serafin Exner, Josef Chmel, Ernst von Feuchtersleben, Karl Littrow, and Haidinger] in a focal point" (Wurzbach 1875, 202), Schmidl's periodical began to publish "daily reports on science and art" that consisted of announcements for lectures, society meetings, or the opening times of imperial collections and other museums.

Born in Bohemia, Schmidl (1802–1863) studied law and philosophy in Vienna and got his first position as a trainee at the Imperial cabinet for coins and antiques. Due to persistent financial problems, Schmidl had to accept several jobs simultaneously. Subsequently, he became adjunct, then temporary substitute professor at the university and an agent of the governmental office for censorship. After four unsuccessful applications for a professorship, he resigned his acceptance as a professor for German language at the lyceum in Verona to serve as a private teacher for the children of Prince Lobkowitz between 1833 and 1844. Using his close contact with state officials, Schmidl was allowed to publish the journal *Österreichische Blätter*, which had around 160 learned subscribers in Vienna. This was a sign that before 1848 learned circles in Vienna, then a city of around 500,000 people, were still comparatively small. After Schmidl had gotten permission to hold extraordinary (public) lectures in geography and aesthetics at the University of Vienna, he worked as an unpaid lecturer for earth study and geography of the Habsburg Empire at the Polytechnic Institute in Vienna, where he taught geography, history, literature, and aesthetics. During the revolution of 1848, Schmidl was appointed member of the Vienna city council and even served for a short time as editor of the

Wiener Zeitung. As a result, his intermediating position between radical political ideas and his high loyalty to the monarchy became a handicap for his career. As he later wrote: "The conservatives have forgotten that I risked my life three times in 1848 and the radicals have not forgotten that I was a key player of the conservatives. Now, everything is lost" (Schmidl 1850a).

Besides his bread-winning occupation as administrator of the Vienna Academy of Sciences, mentioned previously (Sienell 2019), Schmidl became interested in karst landscapes and caves during his extensive travels through the mountainous periphery of the Habsburg Empire. In 1857, he finally accepted a professorship for geography, history, and statistics at the Polytechnic Institute in Budapest, only a few years before he died. Besides his work as an editor, he published around 15 monographs, among them a handbook of geography, literary works, and guidebooks for travelers, flaneurs, and ramblers. With the support of the Academy, he also wrote five books and several articles on different caves of the Habsburg Empire, in which he coined the term "cave study" (*Höhlenkunde*) (Schmidl 1850b), around 40 years before similar terms were developed in other languages.

In contrast to Schmidl, Franz Kraus (1834–1897), who was about 30 years younger and an academically untrained wealthy Viennese businessman, retired at the age of 40 to dedicate himself entirely to the study of natural sciences and caves. He heard lectures at the university and became a member of several associations, such as the Geographical Society, the Alpine Club (*Österreichischer Touristen-Club* or "ÖTC"), and the Scientific Club, where he built up close relationships to well-established scholars like Franz von Hauer, Ferdinand Hochstetter, Josef Szombathy, Eduard Suess, and Felix Karrer. By convincing Hauer and Hochstetter to serve as presidents, Kraus became the driving force behind the foundation of the world's first speleological society (1879–1888). Finally, the association developed into a section for nature study of the ÖTC and gathered people from two cultures of understanding, wealthy townsmen and established academics. Originally planned as an "elite-corps" (Anonymous 1878, 143) of around 50 mainly well-known Viennese scholars that also accepted interested amateurs as members, the society grew quite quickly. In addition, the association stood in close contact with the Natural History Museum, of which Hochstetter was General Director, and the Academy of Sciences' Commission for Prehistory, which was founded shortly thereafter, was also led by Hauer and Hochstetter and would sponsor extensive cave research projects.

The institutionalization of cave study in the form of a private society went hand in hand with the far-reaching changes in knowledge dissemination described previously. Because of the association's position in-between the cultures of science, traveling, and tourism, speleological communities reflected the parallels and conflicts between academic and public views of the earth. However, this form of cooperation had benefits for both parties. On the one hand, established scientists obtained access to data and an audience for their research, while at the same time giving up their exclusive academic prerogative of interpretation; on the other hand, the autodidacts received recognition and institutional or financial support for their work. Subsequently, Kraus was appointed correspondent of the Imperial Geological Survey and volunteer at the Natural History Museum. Besides his articles in alpine yearbooks, journals for popular science, and the 24-volume *Kronprinzenwerk*—a collective portrait of the Austro-Hungarian Monarchy, suggested in 1883 by Crown Prince Rudolf von Habsburg—Kraus finally published one of the first handbooks of cave study in 1894 (Kraus 1891, 1894).

The fact that speleological societies were first established in Vienna can be explained by a specific communicative-spatial setting within the metropolis and between the city and its surroundings, which shaped the field's localization on the cusp of science, politics, the public sphere, and the economy, and also between natural and cultural sciences. On the one hand, Vienna was, and still is, situated at the intersection of the main transportation routes between the limestone mountains of Moravia (north), Hungary/Slovakia (east), Styria/Lower and Upper Austria/Salzburg (west), and Slovenia/Italy (south), where many of the historically known caves are located. On the other hand, as the place where reports of cave explorations and books on cave study were finally published, Vienna—ahead of all other major cities of the Habsburg Empire—became a communication hub for this developing field of research.

If we consider Schmidl's and Kraus's spatial paths through the various (popular) scientific societies in Vienna at a higher level, it becomes clear that both of them did not only retrace their subterranean itineraries. Rather, they told their stories repeatedly to different audiences and used various media. For example, Schmidl reported on his cave ventures during the winter of 1851 not only in the newspaper, but also in the lecture series and printed proceedings of the Academy and the Geological Survey (Schmidl 1851a, 1851b). Therefore, the concentration of their narratives in the urban discourses formed different centers, where people began to get interested in caves and studied them for different purposes: These include the Geological Survey; the Academy of Sciences; the Natural History Museum; the Geographical, Anthropological, Zoological-Botanical, and the Antiquity Societies; diverse alpine clubs; the Scientific Club; and even the Ministries of Agriculture and Commerce, where the economic interest in cave deposits and the use of caves for early tourism purposes developed. Essential for the circulation of go-betweens were the modes of access to these urban communities of knowledge, where the oral and written exchange of these spatial narratives took place. Similar to cave study itself, where questions of publicity and the accessibility of hidden spaces ran right through all itineraries, Schmidl and Kraus needed spaces in-between to enter these societies, whose transitional positions provided free space for professionals as well as amateurs.

That the cooperation among these groups was not free of conflicts is illustrated by a note by Franz Kraus (1880, 7) in one of his travel reports: "Do they really like the contribution of a layman? Who can answer this question? I don't know." Similarly, in a letter sent from Budapest to Vienna, Schmidl (1860) complained about his former position as the head of the Academy's administrative office: "It was nothing else than intellectual slavery. During ten years, these 40 tyrants in the Vienna academy poisoned me, so that I was forced to retreat myself like a polyp in my house of calcite, where I recovered and came back as a man of social and scientific idiosyncrasy. This current state is very convenient for me" (Schmidl 1860).

4. From Spaces in-Between to the Public Sphere: Newspapers and Communicative Strategies

Despite the differences in the biographies of Kraus and Schmidl, there are also clear similarities: Both were prolific writers, were entangled with multiple institutions or fields of knowledge, and could not be exclusively assigned to the group of scientific amateurs or professionals.

Schmidl, who undertook research trips over several months to the karst mountains in Slovenia and Italy between 1850 and 1853, was supported by different institutions, including the newly founded Imperial Geological Survey, the Academy of Sciences, and even some ministries. In particular, Haidinger, for whom Schmidl had published reports and articles in his former journal *Österreichische Blätter*, furnished his expeditions with instruments and sent him as a correspondent of the geological survey to the Slovenian and coastal so-called karst (topography formed by the dissolution of soluble rocks), to explore the opportunity of gaining fertile agricultural land from the often-inundated karstic poljes (large plains) and sinkholes. That Schmidl (1854, VI, VIII) faithfully pursued this arrangement is hard to believe: His "favorite idea" of creating an underground topography of Carniola and passing "most enjoyable hours" on adventurous subterranean journeys might have been stronger. In the first yearbook of the geological survey, Schmidl (1850c, 701) reported from his investigations: "The founding of the Imperial Geological Survey filled me with the hope of realizing my long-lasting plan to investigate the caves of the so-called 'karst' mountains. Thanks to the mediation of the director of this institution, Mr. W. Haidinger, I received support of the most encouraging kind. Equipped with all instruments, I left Vienna on August 9, 1850, to begin my investigations from Planina." To secure the goodwill of his sponsor, Schmidl even named parts of the Planina Cave in Slovenia after Haidinger and reported on his subterranean discoveries in the regular meetings of the Vienna Academy. In addition to five papers on his subterranean ventures printed in the Academy's proceedings, Schmidl finally got permission to publish a two-volume luxury edition (1854) on his researches.

It is particularly noteworthy that Schmidl already used public media to obtain funding, only two years after censorship was lifted. To support his cave investigations through positive campaigning, Schmidl benefited once more from his close relationship to the media. Between 1850 and 1853, he published not less than 21 reports of his research travels in the feuilleton section of the *Wiener Zeitung*. Due to the prominent placement of Schmidl's articles at the beginning of the evening paper, they were also reprinted by other newspapers in different towns of the monarchy, for example the *Laibacher Zeitung*. Additionally, the always-identical title "Aus den Höhlen des Karstes" ("From the karst caves") increased the identifiability of his articles, so that he soon became known throughout Vienna under the nickname *Höhlenschmidl* ("Schmidl from the caves") and was able to create a sustainable public interest in this seemingly bizarre field of research.

A different strategy was followed by Franz Kraus, who could already take advantage of the high number of scientific societies that had been founded in Vienna between 1850 and 1880 (see previous mention). As was the case 30 years before, the argumentation needed to raise public funds for extensive cave research activities was provided by recent newspaper reports of flooding in Slovenia. These periodic incidents, which often afflicted several of the 50 valleys in Carniola at harvest time, meant a severe setback for agriculture in the region and endangered the supply situation of the coastal city of Triest in the long term. In cooperation with Hauer, Kraus advanced Haidinger's idea to gain fertile farmland in the barren coastal karst by draining surface water into subterranean canals and improving muddy or buried suction holes. Two years earlier, Hauer (1883) had already published his karst improvement project in the journal of the Alpine Club and given a lecture on the topic in the premises of the "Scientific Club". In a memorandum to the Central Committee of the "ÖTC" that Kraus published two years later, he demanded to resume immediately Schmidl's former research on underground rivers in order to prove a suspected connection between the caves "Pivka jama" and "Postojnska jama" (now in Slovenia). Despite its institutional affiliation with the "ÖTC", the project was not to be led by that society, but by an intermediating committee that was to bring together scientific, economic, and touristic needs and regularly communicated their results in various scientific journals and newspapers. Especially after the construction of the new Viennese High Spring Water Source Pipeline (*Kaiser-Franz-Josef-Hochquellenwasserleitung*) in 1873, the public interest in questions of water supply and karst hydrology was quite high. As a forum and advisory body for karst and cave research, private interests, research intentions of universities and state museums as well as questions of public utility were discussed and coordinated in the so-called *Karst-Comité* that consisted of scholars such as Hauer, Suess, Szombathy; amateurs such as Kraus; railway and mining officials; politicians; and members of the high nobility as patrons (Szombathy 1885; Hauer 1886).

Although the *Karst-Comité* existed for only two years after its establishment in 1885, when its extensive cave research and karst improvement program were taken over by the Ministry of Agriculture in the person of the forest official and hydrologist Wilhelm Putick, it is worth taking a closer look at Kraus' and the committee's communicative strategies. The "reestablishment of subterranean communications" ("Instandsetzung unterirdischer Communicationen") (Rabl 1897, 31) as the work program of the *Karst-Comité* correlated with its metadisciplinary integration into the research landscape of the city of Vienna. Reports of the *Karst-Comité* were not only printed in single newspapers, proceedings, or journals, but also in various public media. According to the respective communication medium and the public to be addressed, articles were published either by professional scientists or amateurs, anonymous or named.

In summary, Schmidl and Kraus were not only well connected with the media, but also used the strategies of communication and the publicity of Viennese newspapers to support their career goals and projects. For them, Vienna represented both the expeditions' starting point, where they raised research and third-party funds, and a communicative resonating space for addressing different publics.

Conclusion

Like other cities, but to a greater extent, metropolises were distinguished by advanced access to means of communication, and the ability to bind multiple forms of discourse and other entangled spaces of knowledge. Forming communication hubs for multiple publics and fields of knowledge between the sciences and the public sphere can be understood as one defining feature of nineteenth-century metropolises. Major cities like Vienna created their own knowledge-based topographies, where spatial phenomena like concentration, circulation, and transformation took place and processes of increasing urban size, density, and heterogeneity went hand in hand with communicative practices such as collaboration, publication, and translation between different contexts. Consisting as they did of multiple centers and villages, cathedrals and catacombs, global as well as local dimensions of knowledge, metropolises also exhibited fissures, subterranean communication, and transportation highways, where people and objects interacted and circles of exchange were kept intimate. As is the case at airports and train stations today, spaces in-between and ephemeral transitional zones of knowledge were populated by different urban forms of life as merchants, travelers, mediators, seducers, onlookers, and 'homeless' researchers. Building bridges and circulating between mutable fields and cultures of knowledge, go-betweens, mediators, and objects of scientific importance played a key role for the concentration of knowledge in major urban environments.

In the case of Vienna, the impact of scientific societies on the production and dissemination of knowledge should neither be underestimated nor undervalued in comparison with state-run research establishments. Especially during the bourgeois era and even before 1848, when new forms of scientific communication and publicity gradually developed, learned societies represented not only venues for scientific and political negotiation processes, but also symbolized elements of continuity during processes of political and urban transformation, especially in conjunction with the architectural design and construction of the monumental Ringstrasse. In the second half of the nineteenth century, scholarly associations in Vienna embodied the publicity and orality of scientific activities and subsequently became the birthplace of new fields of research such as cave study, the proponents of which used public communicative strategies to reach their goals. In this period, natural scientists and the broader urban middle class began to question the academic knowledge monopoly and searched for new forms of scientific communities and audience-orientated knowledge transfer. Establishing alternative cultures of meta- and transdisciplinary understanding, they favored a holistic approach in the tradition of Alexander von Humboldt's book *Kosmos* (1845–1858) instead of relying on the acquisition of detailed knowledge and scientific specialization. The boom of (popular) scientific societies in Vienna lasted until the beginning of the 1920s, when political radicalization, the effects of inflation, and the separation from other learned societies in the former Habsburg Empire forced the development of new exclusive and alternative intellectual circles that were no longer primarily interested in publicity or in a holistic approach to science.

Notes

1. The Academy of Sciences in Vienna, originally intended to be an empire-wide institution, had a special status. From the end of the 18th century until 1867, the abbreviation *k. k.* or *kaiserlich-königlich* was used for empire-wide state institutions. The first *k.* (*imperial*) stood for the title of the Roman-German Emperor, from 1804 for the title of the Emperor of Austria. The second *k.* (*royal*) stood for the title of the King of Hungary. In the founding patent of Emperor Ferdinand, printed in the *Wiener Zeitung* on 17 May 1847, the academy was referred to as a "k. k. Akademie" (*Imperial-Royal Academy*). In the following years, the academy might have changed its name on its own initiative. In the first published almanac of the academy (1851), where its statutes were printed, the name "Imperial Academy of Sciences" was already used. This corresponded to the fact that in the 1850s many Hungarian members, originally nominated by the emperor, died or resigned their membership. Instead, many Viennese scholars were voted as new members. Similarly, it is very likely that the academy wanted to emphasize its autonomy by avoiding the title of a state institution.

2. Although today the titles *k. k.* and *k. u. k.* are often used as synonyms, both terms are historically and legally distinct (see endnote 1). After the Austro-Hungarian Compromise of 1867, the abbreviation *k. k.* was used only for state institutions in the Austrian part of the monarchy and referred to the Empire of Austria and the Kingdom of Bohemia. Instead, in the dual monarchy the abbreviation *k. u. k.* (*imperial and royal*) was used to describe common institutions of both halves of the empire. Nevertheless, many *k. k.* institutions such as the Geological Survey, the Natural History Museum or the Geographical Society, which were founded before the Austro-Hungarian Compromise, kept their titles. This is due to the fact that after 1867 corresponding institutions were established in the Hungarian half of the monarchy. In some cases, such as the Military Geographic Institute (1889), the title was subsequently changed from *k. k.* to *k. u. k.*

Bibliography

Anonymous [probably Lichtenstern, J. v.] (1810). *Kurze Nachricht von der Verfassung und den Beschäftigungen des Cosmographischen Instituts in Wien seit seinem Anfange bis zum Jahre 1810.* Vienna: self-published.

Anonymous (1847a). Chemie, Topographie, Mineralogie, Physik, Paläontologie, Geologie. *Wiener Zeitung* 344, 1686.

Anonymous [probably Schmidl, A.] (1847b). Versammlungen der Freunde der Naturwissenschaften. Bericht vom 26. November. *Österreichische Blätter für Literatur, Kunst, Geschichte, Geografie, Statistik und Naturkunde* 294, 1166–1168.

Anonymous [probably Schmidl, A.] (1847c). Die kaiserl. Akademie der Wissens. in Wien. *Österreichische Blätter für Literatur, Kunst, Geschichte, Geografie, Statistik und Naturkunde* 2074, 821–824.

Anonymous (1848). Wo sind die Schriftsteller Wiens? Warum treten sie nicht zusammen, um das Pressegesetz zu beraten und die Akademie der Wissenschaften? *Wiener Abendzeitung* 7, 27.

Anonymous [Schmidl, A.] (1849). Die kaiserliche Akademie der Wissenschaften: Part I, II & III. *Beilage zur Allgemeinen Zeitung (Augsburg)* 127, 179, 224, 1957–1958, 2768–2769, 3464–3466.

Anonymous (1878). Für Höhlenfreunde. *Neue deutsche Alpenzeitung* 12, 143.

Ashcroft, B., Griffiths, G., and Tiffin, H. (2000). *Postcolonial Studies: The Key Concepts.* London and New York: Routledge.

Barták, J., Hrdina, I., Romancov, G., and Zlámal, J. (eds.) (2007). *Underground Space: The Fourth Dimension of Metropolises.* London: Taylor & Francis.

Boué, A. (1869). *Ein freies Wort über die kaiserliche Akademie der Wissenschaften sammt Vergleich der Akademien mit den freien, gelehrten Vereinen.* Vienna: Wilhelm Braumüller.

Brauneder, W. (1992). *Leseverein und Rechtskultur. Der Juridisch-politische Leseverein zu Wien 1840 bis 1990.* Vienna: Manz.

Coen, D. R. (2018). *Climate in Motion. Science, Empire, and the Problem of Scale.* London, Chicago: University of Chicago Press.

Csendes, P. (2003). Stadt und Technik. Wissenschaftlicher Fortschritt und urbane Entwicklung. In: Csendes, P., and Sipos, A. (eds.), *Budapest und Wien. Technischer Fortschritt und urbaner Aufschwung im 19. Jahrhundert.* Budapest, Vienna: Franz Deuticke, 5–18.

Daum, A. W. (1998). *Wissenschaftspopularisierung im 19. Jahrhundert. Bürgerliche Kultur, naturwissenschaftliche Bildung und die deutsche Öffentlichkeit, 1848–1914.* Munich: Oldenbourg.

Dotzler, B. J., and Schmidgen, H. (2008). Einleitung. In: *Parasiten und Sirenen. Zwischenräume als Orte der materiellen Wissensproduktion.* Bielefeld: Transcript, 7–17.

Felt, U. (2000). Die Stadt als verdichteter Raum der Begegnung zwischen Wissenschaft und Öffentlichkeit. Reflexionen zu einem Vergleich der Wissenschaftspopularisierung in Wien und Berlin. In: Goschler, C. (ed.), *Wissenschaft und Öffentlichkeit in Berlin, 1830–1930.* Stuttgart: Franz Steiner, 185–220.

Felt, U. (2002). Wissenschaft und Öffentlichkeit—Wechselwirkungen und Grenzverschiebungen. In: Ash, M. G., and Stifter, C. (eds.), *Wissenschaft, Politik und Öffentlichkeit. Von der Wiener Moderne bis zur Gegenwart.* Vienna: Wiener Universitätsverlag, 47–72.

Fötterle, F. (1868). Rechenschafts-Bericht des ersten Sekretärs. *Mittheilungen der k. k. Geographischen Gesellschaft* 12, 27–29.

Frauenfeld, G. v. (1852). Gründende Versammlung am 9. April 1851. *Verhandlungen des Zoologisch-Botanischen Vereins in Wien* 1, 1–3.

Frauenfeld, G. v. (1863). Bericht des Secretärs. Jahressitzung am 10. April 1863. *Verhandlungen der k. k. Zoologisch-Botanischen Gesellschaft in Wien* 12, 29–31.

Habermas, J. (1989). *The Structural Transition of the Public Sphere: An Inquiry into a Category of Bourgeois Society.* Trans. T. Burger, and F. Lawrence. London: Polity Press. First published 1962.

Haidinger, W. (1869). *Das k. k. Montanistische Museum und die Freunde der Naturwissenschaften in den Jahren 1840 bis 1850.* Vienna: Holzhausen.

Hammerschmidt (Abdullah), C. E. (1848). Vortrag von Herrn Dr. Hammerschmidt. *Berichte über die Mittheilungen von Freunden der Naturwissenschaften in Wien* 4, 1–6, 274–277.

Hartwig, G. (1871). *The Subterranean World.* New York: Scribner & Welford.

Hauer, F. v. (1883). Berichte über die Wasserverhältnisse in den Kesselthälern von Krain. *Österreichische Touristen-Zeitung,* 3–4, 25–31, 37–41.

Hauer, F. v. (1886). Jahresbericht für 1885. *Annalen des k. k. Naturhistorischen Hofmuseums* 1, 1–46.

Humboldt, A. v. (1845–1858). *Kosmos—Entwurf einer physischen Weltbeschreibung* (4 Vols.). Stuttgart & Augsburg: Cotta.

Hye, H. P. (1988). Vereinswesen und bürgerliche Gesellschaft in Österreich. *Beiträge zur historischen Sozialkunde* 19, 86–96.

Hye, H. P. (1992). Zur Liberalisierung des Vereinsrechtes in Österreich. Die Entwicklung des Vereinsgesetzes von 1867. *Zeitschrift für Neuere Rechtsgeschichte* 14, 191–216.

Imperial Academy of Sciences (1847). Protocol of the Academy-Meetings (both divisions) on the 18th December 1847. Archive of the Austrian Academy of Sciences.

k. k. Statistische Central-Commission (1892). *Handbuch der Vereine der im Reichsrathe vertretenen Königreiche und Länder.* Vienna: Manz.

Kadletz-Schöffel, H. (1992). *Metternich und die Wissenschaften* (2 Vols.). Vienna: Verband Wissenschaftlicher Gesellschaften Österreichs.

Kanzlei des Wissenschaftlichen Club (1876). *Der "Wissenschaftliche Club". Kurze Darstellung seines Entstehens und seiner Hilfsmittel.* Vienna: self-published.

Kittler, J. (2013). The City. In: Simonson, P., Peck, J., Craig, R. T., and Jackson, J. P. (eds.), *The Handbook of Communication History*. New York and London: Routledge, 273–288.

Klemun, M. (1998). *Werkstatt Natur. Pioniere der Forschung in Kärnten. Katalog zur Ausstellung des 150jährigen Bestehens des Naturwissenschaftlichen Vereines für Kärnten*. Klagenfurt: Verlag des Naturwissenschaftlichen Vereines für Kärnten.

Klemun, M. (2003). Der Siebenbürgische Verein für Naturwissenschaften zu Hermannstadt im Netzwerk der Habsburgermonarchie. In: Heltmann, H., and Killyen, H. v. (eds.), *Der Siebenbürgische Verein für Naturwissenschaften zu Hermannstadt (1849–1949)*. Sibiu, Heidelberg: Hora & Arbeitskreis für Siebenbürgische Landeskunde, 35–46.

Klemun, M. (ed.) (2020). *Wissenschaft als Kommunikation in der Metropole Wien. Die Tagebücher Franz von Hauers der Jahre 1860–1868*. Vienna, Cologne, and Weimar: Böhlau.

Kraus, F. (1880). Höhlenfahrten. *Literatur-Anzeiger* (Club for Cave Study) 5, 4–8.

Kraus, F. (1891). Der Karst. In: Habsburg, R. v. (ed.), *Die österreichisch-ungarische Monarchie in Wort und Bild* ("Kronprinzenwerk", Vol. 8). Kärnten und Krain. Vienna: Hof- und Staatsdruckerei, 285–304.

Kraus, F. (1894). *Höhlenkunde. Wege und Zweck der Erforschung unterirdischer Räume. Mit Berücksichtigung der geographischen, geologischen, physikalischen, anthropologischen und technischen Verhältnisse*. Vienna: Gerold.

Lefebvre, H. (1992). *The Production of Space*. Edited and Translated by D. Nicholson-Smith. Malden: Blackwell.

Mattes, J. (2015). *Reisen ins Unterirdische. Eine Kulturgeschichte der Höhlenforschung*. Vienna, Cologne and Weimar: Böhlau.

Mattes, J. (2019). *Wissenskulturen des Subterranen. Vermittler im Spannungsfeld zwischen Wissenschaft und Öffentlichkeit*. Vienna, Cologne, and Weimar: Böhlau.

Nieto-Galán, A. (2016). *Science in the Public Sphere*. New York and London: Routledge.

Pike, D. L. (2005). *Subterranean Cities: The World Beneath Paris and London, 1800–1945*. Ithaca: Cornell University Press.

Pusman, K. (2008). *Die "Wissenschaften vom Menschen" auf Wiener Boden (1870–1959). Die Anthropologische Gesellschaft in Wien und die anthropologischen Disziplinen im Fokus von Wissenschaftsgeschichte, Wissenschafts- und Verdrängungspolitik*. Vienna, Berlin: LIT.

Rabl, J. (1897). † Franz Kraus. *Österreichische Touristenzeitung* 17:3, 30–31.

Raj, K. (2016). Go-Betweens, Travelers, and Cultural Translators. In: Lightman, B. (ed.), *A Companion to the History of Science*. Malden and Oxford: Wiley, 39–57.

Ranzmaier, I. (2013). *Die Anthropologische Gesellschaft in Wien und die akademische Etablierung anthropologischer Disziplinen an der Universität Wien, 1870–1930*. Vienna, Cologne, and Weimar: Böhlau.

Raptis, K. (1998). Der wissenschaftliche Klub 1875–1900. Studie zur Frühzeit eines bürgerlichen Vereins in Wien. *Wiener Geschichtsblätter* 53, 38–59.

Schmidl, A. (1847). Die Elemente des geistigen Lebens in Wien. *Österreichische Blätter für Literatur, Kunst, Geschichte, Geografie, Statistik und Naturkunde* 4:56, 221–224.

Schmidl, A. (1850a). Letter from Schmidl to Josef Feil (2.4.1850). Manuscript Collection, Legacy of Josef Feil. Vienna City Library (H.I.N.-129783).

Schmidl, A. (1850b). Beitrag zur Höhlenkunde des Karst. *Sitzungsberichte der kaiserlichen Akademie der Wissenschaften in Wien, mathematisch-naturwissenschaftliche Klasse 5*, 446–479.

Schmidl, A. (1850c). Die Untersuchung einiger Höhlen im Karst. *Jahrbuch der k. k. Geologischen Reichsanstalt 1*, 701–705.

Schmidl, A. (1851a). Ueber den unterirdischen Lauf der Recca. *Sitzungsberichte der kaiserlichen Akademie der Wissenschaften in Wien, mathematisch-naturwissenschaftliche Klasse 6*, 655–662.

Schmidl, A. (1851b). Uebersicht der im vorigen Winter im Auftrage des hohen k. k. Handelsministeriums ausgeführten Untersuchungen über den unterirdischen Lauf der Recca. *Jahrbuch der k. k. Geologischen Reichsanstalt 2:2*, 184–185.

Schmidl, A. (1852). *Wien und seine nächsten Umgebungen: mit besonderer Berücksichtigung wissenschaftlicher Anstalten und Sammlungen* (5th ed.). Vienna: Gerold.

Schmidl, A. (1854). *Zur Höhlenkunde des Karstes. Die Grotten und Höhlen von Adelsberg, Lueg, Planina und Laas*. Vienna: Braumüller.

Schmidl, A. (1860). Letter from Schmidl to Ludwig August Frankl von Hochwart (17.10.1860). Manuscript Collection. Vienna City Library (H.I.N.-103980).

Sienell, S. (2019). *Das Verwaltungs- und Dienstpersonal der Akademie der Wissenschaften 1847 bis 1960. Eine kleine Sozialgeschichte*. Vienna: Austrian Academy of Sciences.

Stubenrauch, M. v. (1857). *Statistische Darstellung des Vereinswesens im Kaiserthume Österreich*. Vienna: Hof- und Staatsdruckerei.

Szombathy, J. (1885). Die bisherige Thätigkeit des Karst-Comités des österreichischen Touristen-Clubs. *Mittheilungen der Section für Höhlenkunde des Ö.T.C. 2*, 17–20.

Tietze, E. (1907). Ansprache des Präsidenten. *Mitteilungen der k. k. Geographischen Gesellschaft 50*, 76–85.

Verne, J. (1864). *Voyage au Centre de la Terre*. Paris: Hetzel.

Wurzbach, Constant v. (ed.) (1875). *Biographisches Lexikon des Kaiserthums Oesterreichs* (Vol. 30). Vienna: Hof- und Staatsdruckerei.

10 Science-Oriented Popular Education

Heterotopic Learning Venues for Scientific Knowledge in Vienna, 1887–1918

Christian H. Stifter

Introduction: Heterotopic Counterplaces— "Localized Utopias"

The reflection on social spaces, the specific otherness of which regarding their role, function, and use places them in a kind of antipodal counter-relationship to other contemporary places, is of comparatively recent date. In his 'heterotopology,' Michel Foucault developed the concept of 'counter-spaces' or 'localised utopias' in 1966. His work provides a theoretical approach to special places, where several actually incompatible spaces of action or practices intertwine—a topological-praxeological superposition or penetration that questions or negates all other socially existing spaces. As counterpoints to other spaces of action and experience in society, imaginations realized in such a limited way produce "another quite real space" that, as Foucault put it, "has a perfect order" of its own (Foucault 2017, 19–20).

These 'counterplaces' are indeed connected with other everyday living and action spaces, but in the form of a distinctly contradictory and radical relationship. In contrast to utopias, which have no real location, heterotopias are real places, which, however, always have a "system of opening and closing off which isolates them from their surroundings" (ibid., 18). Through their spatio-temporal extraterritoriality and their specific exclusivity as "real places beyond all places" (ibid., 11), they are able to leverage, invert, or cancel social relations.

Heterotopias—Foucault cites as examples cemeteries, retirement homes, gardens, mental hospitals, prisons, holiday villages, zoological and botanical gardens, as well as museums, libraries, and theatres—would in this sense be "completely different spaces" (ibid.) in which "heterochrony", a kind of counter-time experience, takes place that shows a changed direction toward and deviating speed from the development of society as a whole—an alternative dimension of experience and opposing perception of temporality and progression, the utopian potentials of which "contradict" ordinary experience and sometimes silence it (Defert

2017, 75). Such places might be, for instance, a 'festival' of equal learning and free teaching, of collectively designed intellectual developments that seemingly stand outside everyday, limiting power relations and constellations. Such places of learning could establish, so to speak, a social and discursive space of empowerment, which, in its specific social formation, is both an intellectual space of discourse and a specific space of social action of the kind described by Henri Lefebvre (1991, 25–26).

Building on this theoretical concept of an asynchronous, resistive, isolated cultural 'counterspace', the following chapter will examine Viennese learning locations where a novel model of an urban, science-oriented form of knowledge transfer for adults developed toward the end of the nineteenth century. These places, created out of a late-Enlightenment liberal spirit and a civil society reform mentality, were both manifest expressions of new approaches to teaching and learning in specially created new spaces and the result of new social alliances in which different, partly diametrically opposed social groups of actors and political and ideological positions cooperated.

1. Proponents, Positions, and Programmatics of Viennese Popular Education Around 1900

The cultural and social dimensions of this first large-scale popular education offensive in the metropolitan area of Vienna are reflected—apart from the content of the program—not least in the commitment of its central sponsors and promoters, most of whom came from the intellectual and cultural elite. In its effects, this civil society initiative for a broad-based mediation of education and scientific knowledge is to be understood both as a reaction to the great social, cultural, and political upheavals of the nineteenth century and as an independent transformative force, as is also the case for changes in the fields of science, politics, and economics during this period.

The processes of industrialization and urbanization, which began in the Austro-Hungarian Monarchy somewhat later than in other European regimes, led—against the background of increasing demands for political participation—to lasting changes in the social, economic, and political-administrative spheres, as well as in the structure of traditional approaches and forms of transmission of knowledge and education. Driven by the numerous achievements of modern natural science and technology, which, from the point of view of scientific research, resulted in a veritable "orgy of scientific triumphs" (Whitehead 1988, 122), to which both the hopes for progress of the liberal bourgeoisie and the social democratic labor movement were tied, the social democratic labor movement chose to take a new approach from the middle of the century onwards. In addition to the existing political power relations throughout Europe, traditional educational barriers on a broad front were called into question, barriers which were of particular political relevance in the metropolitan area, in view of enormous urban population growth, with all its social downsides and inequalities.

As the leading social democrat Otto Bauer wrote in 1905, the educational consensus and unity of Viennese popular education pioneers established, especially in the period following the liberal regime of the 1870s, a counterposition to the reactionary Christian social city regime, which "united men of progress of all parties in opposition" (Bauer 1905, 462). In his pointed analysis, Bauer stated: "The intelligentsia hates clericalism, and the bourgeoisie fears guild-like agrarian reaction even more than the working-class movement. Increased popular education appeared to both as a means to combat dreaded cultural and economic-political reactionary aspirations" (ibid.).

The broad homogeneity of the content and programmatic orientation of the Viennese adult education movement was based on two central positions. One was vehement criticism of the public school system, from which an urgent need for compensatory action was derived from both an educational and a democratic perspective. Despite the school reform of 1869, the illiteracy rate was 25 percent in Vienna before the First World War, not least due to the enormously widespread use of child labor at school age (38 percent) (Rumpler and Seeger 2010, 228–229). Although improvements took place in the professional quality of teaching staff, persisting authoritarian-paternalistic learning and drill-like teaching methods were sharply criticized, especially in elementary education. School pedagogy had by no means completely detached itself from the "k.k." school regulations for the Habsburg hereditary states from 1806 (Engelbrecht 1984, 521–522) designed to produce "quite cordially good, direct, and busy people", in a word, Catholics loyal to the empire (Ehalt 1978, 123).

In addition, criticism was also directed at the secondary school system, which was entrusted with the preparation for university studies in the wake of the anti-liberal-etatistic Thun-Hohenstein university reform of 1849 and its establishment of the Matura examination (formally equivalent to the German *Abitur*) as the sole access route to higher education. In his *Popular Scientific Lectures* published in 1896, the physicist and philosopher of science Ernst Mach criticized the lack of modern general school education, including the low importance of scientific and mathematical-technical subjects. Like Georg Christoph Lichtenberg before him, he also feared that the function of the schools would be "simply to select the persons best fitted for being drilled, whilst precisely the finest special talents, which do not submit to indiscriminate discipline, would be excluded from the contest" (Mach 1896, 330). In particular, Mach denounced the accumulation of undigested knowledge, as well as the "rushing out and mere proclamation" (ibid., 326) of content as applied educational methods, because "uniforms are excellent for soldiers, but they are not suitable for the head" (ibid., 329).

For the time being, however, the realization of progressive demands in terms of content and didactics in the school system was limited to private initiatives, whose founders, such as the members of the anticlerical association "Free School" (*Freie Schule*) founded in 1905, were close to science-oriented popular education or themselves among its leading protagonists, for example the deputy chairman of the association, medieval historian Ludo Moritz Hartmann (Stifter 2015a, 247–263).

The second critical position taken by the representatives of science-oriented popular education was its opposition to the traditionally elitist corporate structure of the university, access to which, as the German social democrat Wilhelm Liebknecht argued in a programmatic speech in 1872, was completely closed to ordinary working people (Liebknecht 2012, 154). In view of the methodologically and didactically backward practice of university teaching and its isolation from society, reform-oriented university lecturers and professors desired modernization and demanded a new connection between science and the public, in the form of a curriculum accessible to the general population, the contents of which were to be presented in understandable language.

According to the "exotericism" of scientific knowledge demanded by the medievalist, educational reformer, and outstanding pioneer of adult education, Ludo Moritz Hartmann, rays of light should emanate from the "focal point of science" and "illuminate the exoteric circle of the population" (quoted from Kapner 1961, 9). This hope was also shared by his comrades-in-arms. One such like-minded thinker was the chairman of the Vienna Popular Education Association (*Volksbildungsverein*) and professor of philosophy at the University of Vienna, Friedrich Jodl (Frank 1970). With reference to Friedrich Schiller's writings on social and historical sciences, he said that "science and art . . . must not remain the sole property of the upper social strata if a prosperous development of the entire population is to take place" (Jodl 1912, 1). Aesthetic and spiritual enlightenment were seen here as an enabling space for political freedom, as a "prerequisite of the reasonable state". However, it is important to stress that Jodl understood the interrelation quite dialectically. Writing two years before the outbreak of the First World War, he stated not only that the people need "contact with science" in order to orient themselves in life and to expand their view of business and day-to-day work, but also that science could not "in a democratic age do without the stability and the support that respect for science and understanding of its life-promoting power grant" (ibid.).

Hopes for progress of this kind were supported by technical achievements and medical innovations, which clearly demonstrated the power of modern science to change society and were not based on a naive view of the real state of universities. Thinking in terms of modern, empirically founded science, which Mach explicitly claimed to be of equal rank with the everyday use of reason (Mach 1896, 2), functioned as an ideal-typical orientation model for critical reflection.

After the containment of church power following the abrogation of the Concordat with Rome in 1870, the University of Vienna experienced an upswing during the brief phase of liberal rule. In addition to a significant increase in the number of students, not least due to the increased demand for qualified personnel in administration, the spread of positivistic natural science also led to a surge in modernization, which was reflected in a number of outstanding achievements in the individual sciences (Feichtinger 2010). Nevertheless, parallel to Christian social reconfessionalization efforts from the early 1890s onwards, the university became a place of national, anti-Semitic and violent conflict (Rathkolb 2013; Seebacher 2010). In addition, traditional didactic standards of university teaching, which were still only accessible to a small social minority, had hardly changed (Stifter 2015b, 302–303).

In her study "Imperial Vienna", published in 1905, the English cultural and art historian Amalia Sarah Levetus, who moved from Birmingham to Vienna in 1891, briefly described then-current teaching practices at the University of Vienna against the background of her own academic socialization at the universities of Cambridge, Birmingham, and Aberdeen (Filla 2001). In 1897, at the invitation of national economist Eugen Schwiedland, she became the first woman to give lectures at the University of Vienna, on economic cooperatives of English and Scottish workers (Vortrag einer jungen Dame an der Universität Wien 1897).

Through her regular involvement in a number of associations and institutions, her active participation in the life of Viennese society, as well as her journalistic activities, "Miss Levetus" was socially well connected. For example, she was a member of the "Association for the organization of academic lectures for ladies" and the *Wiener Frauenclub*, founded in 1900 (*Sport & Salon* 1905), and a friend of women's rights activist Marianne Hainisch. In addition, Levetus ran an English-language institute and worked as a permanent Vienna correspondent of the leading London Art Nouveau magazine *The Studio* (Large 2003).

In contrast to university life in England, Levetus said—apart from the lectures given *ex cathedra*—there was absolutely no social contact between professors and students in Vienna, neither in public nor in private. The author regretted this, since due to the numerous and ethnically diverse audiences "social intercourse would be the only means that the professor would have of becoming acquainted with his students, a mutual advantage to both teachers and taught" (Levetus 1905, 350).

On the other hand, Levetus paid great respect to Viennese popular educational institutions, with which she herself had been closely associated for more than three decades as an English-language teacher and director of the "John Ruskin Club" at the Volkshochschule *Volksheim* in Ottakring (discussed in detail later on). Levetus described the specifics of those institutions, to which she attributed a continental leadership role, as follows: "All languages, science, history, in fact everything which is taught at the university, can be learned here for the modest sum of five crowns. . . . The professor makes excursions to points of interest with the students, social intercourse is promoted, and everything is done *to bring about good-will between teachers and students.* Both men and women attend the lectures and show great eagerness and attention in learning" (ibid. 354; emphasis in the original). According to all available sources, the picture of a high-quality teaching organization in which top-class scholars and students of different social backgrounds and all age groups engaged in an open and benevolent social exchange, undoubtedly drawn here with great sympathy, corresponded perfectly to reality. In the following sections this approach will be described in greater detail.

2. Key Characteristics of the Popular Education Institutes in the Imperial Metropolis Vienna

After isolated initiatives to popularize science and the founding of a number of municipal popular education associations in the Cisleithanian lands, for example in Graz (1870), Linz (1872), and Krems (1885), the boom in popular education that began in most European states at the end of the 1880s also reached the imperial metropolis of Vienna. Within a short period of time, popular education institutions oriented toward scientific knowledge and rationality were founded before the beginning of the twentieth century, the practices of which set a trend for further developments. In particular, these were the Vienna Popular Education Association (*Wiener Volksbildungsverein*), founded in 1887, mentioned previously; the Popular University Lectures (*Volkstümliche Universitätsvorträge*), established in 1895 by the Academic Senate of the University of Vienna on the model of the Universities of Oxford and Cambridge (Taschwer 2002); the Vienna Urania, founded in 1897 with the support of the Lower Austrian Reform Club; and finally, the first European 'People's University' (*Volksuniversität*), founded in 1901 in the Vienna working-class district of Ottakring (Stifter 2005).

The special feature of these places of learning, supported by both the liberal bourgeoisie and the aspiring social democratic labor movement, was a twofold approach. In the first place was a novel cooperation of modern science and potentially all nationalities on an institutional basis. The term 'people' or 'popular', traditionally used in a rather derogatory manner in the context of education and upbringing until the nineteenth century, now defined the clientele mainly in a non-nationalistic, social respect. The target group addressed in Vienna was very broad, including potentially all interested inhabitants, irrespective of their particular social origin, property ownership or income situation, previous knowledge, religious convictions, or gender. The postulate of equality and the moral-philosophical universalism of the Enlightenment were intertwined here with the social and educational model of a democratic public sphere—which, like universal suffrage, had yet to be established. Ludo Moritz Hartmann in particular tried to underline this connection between science-oriented popular education and democracy by pointing out that it was "no accidental coincidence" that the building of the First People's University was opened "on the same day [November 5, 1905] on which the first large demonstration in the struggle for universal suffrage filled the streets of Vienna" (Hartmann 1921, 2).

From the point of view of social reform, this egalitarian principle applied to people's education meant above all assistance for the socially underprivileged classes. Science-oriented Viennese popular education was provided to a successively increasing proportion of workers among the participants; by the 1920s, this was on average 40 percent (Filla 1994, 2–3). The integrative, emancipatory, and participatory setting created a socio-culturally diverse educationally enabling space, the concept and structure of which was based on the principle of equal status for all actors involved. Analogous to the previous demolition of the medieval city walls, including the undeveloped Glacis (the stretch of empty land immediately outside the walls), and the incorporation of the Viennese suburbs, the democratization of access to science, education, art, and literature, starting from local crystallization points, was to make a profoundly inclusive, socially integrative contribution to the "intellectual expansion of the city" (Jodl 1911, 4).

On the part of the authorities, in particular the Catholic clergy and the conservative anti-Semitic, German nationalist, and German *völkisch* press, respectively, facilitating access to education and science for the broad mass of the working population was regarded as a dangerous effort to undermine the existing social order and was thus vehemently rejected, not least because one suspected either a Masonic or a Jewish conspiracy. For example, the *Reichspost*, a conservative newspaper, attacked this liberal "deformation of true popular education", allegedly aimed at injecting an anti-Christian, "Jewish-Liberal plague bacillus" into the population (*Reichspost* 1900, 54).

Besides the connection between science and the people, the second key aspect of science-oriented popular education lay in the late Enlightenment maxim to which every extensive popular education work aimed at mass audiences, but also every intensive work in popular education was then and still remains committed: namely, the advancement of rational, informed, independent, and critical thinking, the ideal type of impartial, ideologically free, and undogmatic, i.e., "unconditional science" that Ludo Hartmann constantly demanded (Hartmann 1920, 132). In the fight against trash literature, alcohol abuse, and social misery, the democratization of access to science and education was seen as a "social duty" to immunize all citizens against vice, mental narrowness, and "any authoritarian worldview", and at the same time to achieve the social betterment of the working class (Hartmann 1901, 165).

This goal was to be achieved by the strict decoupling of knowledge transfer and ideology. The intention was to assure the objectivity and political neutrality of the educational program, which was to be kept free of all day-to-day, party-political questions, and by the professional quality of the lecturers (Ganglbauer 1999). These came preferably from university research, since only specialized university teachers—once they had learned to speak freely and 'popularly'—would be sufficiently close to their subject and, in the best case, "above the subject matter". But that was not all, Hartmann continued: "the best researcher, the best teacher, and the best speaker would be just good enough for the tremendously difficult task they are given," although he acknowledged that this combination would be rare (Hartmann 1910, 117).

As a result of this successfully implemented scientific orientation, the promotion of critical, empirically based reflection, good material resources, and broad social support, the Viennese School of Popular Education, whose protagonists were intensively networked with one another both locally and internationally, achieved attention and recognition in Europe and beyond. For example, in her comparative study on labor education in Vienna and Berlin, published in 1911, Hertha Siemering, head of the Department for Female Youth Care at the Berlin Central Office for Public Welfare, clearly saw the imperial city of Vienna as a "pioneer (that) had preceded all German cities in the Reich" and as the center of the German popular education system (Siemering 1911, 5). She came to the conclusion that, unlike Vienna, all efforts in Berlin were marked by disunity and fragmentation, and that due to the frosty, hostile relationship between social democracy and the bourgeois liberals, "any official community of bourgeoisie and 'proletariat' . . . seemed impossible" (ibid., 71). Moreover, all state institutions in Prussia were separated from official socialist workers' organizations by a deep divide. Not least for this reason, in contrast to Vienna, it had not been possible in Berlin to anchor popular university lectures at the university; this succeeded there—albeit to a much more modest degree—only with the aid of purely private organizations.

The fact that "Vienna was a particularly suitable ground" for the development of an anti-metaphysical scientific world view was later noted in the opening chapter of the Manifesto of the 'Vienna Circle,' published in 1929 (Verein Ernst Mach 1929, 6). Under the influence of political liberalism, Enlightenment ideas, empiricism, utilitarianism, and the "anti-metaphysical spirit" were cultivated above all by world-renowned scholars such as Theodor Gomperz, Eduard Suess, Friedrich Jodl, Ernst Mach, Ludwig Boltzmann, and Adolf Stöhr (*Wissenschaftliche Weltauffassung* 1929, 301). Furthermore, the manifesto emphasized, it is thanks to this "spirit of the Enlightenment" that "Vienna was leading in scientifically oriented *popular education*," exemplified by the foundation of the Vienna *Volksbildungsverein* by Friedrich Jodl as well as by the foundation of the "popular university lectures" and the *Volksheim* by Ludo Hartmann, "whose anti-metaphysical attitude and materialistic conception of history was expressed in all his work" (ibid.). Of the 19 core members of the Vienna Circle, 11 appeared as lecturers at Viennese adult education centers, including 5 longtime popular university instructors such as Hans Hahn, Philipp Frank, and Viktor Kraft. In addition, central pioneers of the scientific world view, such as the aforementioned Eduard Suess, Friedrich Jodl, Ernst Mach, Ludwig Boltzmann, and Adolf Stöhr, were among the proponents and, in some cases, active comrades-in-arms of science-oriented popular education, along with a number of other university teachers (Stadler 1997).

3. The *Volksheim* Ottakring—the First "People's University" in the Habsburg Monarchy

A special field of experimentation and a laboratory of science-oriented cooperation between experts and laymen, as well as alternative forms of teaching and learning in an egalitarian and socially open setting, was the first continental European 'People's University' (*Volksuniversität*) which emerged directly from the popular university lectures of the University of Vienna and whose *genius loci* I will now examine in more detail (Figure 10.1).

Inspired by the listeners of the popular university lecture in philosophy given by university professor Adolf Stöhr, who expressed the wish for a more intensive and concentrated discussion at a central location—the lecture series of the University Extension all took place outside the university building—the opening meeting of the new association for popular education took place on February 24, 1901, in the ballroom of the Vienna Ronacher.

By the end of 1900, 64 persons had signed the founding appeal. Ranging from Habsburg conservatives and liberal Catholics to German liberals and anticlerical social democrats, the signatories illustrated the initiative's broad social support. They included university professors and lecturers such as Karl Brockhausen, Max von Gruber, Edmund Bernatzik, Friedrich Jodl, Anton Lampa, Ernst Mach, Emil Zuckerkandel, Richard von Wettstein, Adolf Stöhr, Julius Tandler, Albrecht Penck, Eduard Suess, Heinrich Swoboda, Eugen von Philippovich, Gustav von Escherich, Friedrich Hertz, Gustav Seidler, Walter Schiff, Laurenz Müllner, and Eduard Reyer. In addition to chairmen of social democratic workers' education associations, members of the Reich Council, the upper house of the empire's parliament, such as Engelbert Pernerstorfer, Julius Ofner, and Karl Seitz also signed the founding appeal. Further supporters were educational reformers such as Else Federn, Carl Furtmüller, Michael Hainisch, women's rights activist Rosa Mayreder, artists such as Tina Blau and Georg Reimers, as well as writers such as Marie von Ebner-Eschenbach, Marie Eugenie delle Grazie, and Ferdinand von Saar (Aufruf zur Gründung 1901, 3).

Figure 10.1 Exterior view of the new building of the 'People's University' (*Volks-hochschule "Volksheim"*) in the sixteenth district of Vienna, completed in 1905. Characteristic of the design was the abandonment of ornamental decoration and representational formulas of the soberly designed façade, in the formal language of Otto Wagner. The attic at the edge of the roof shows a stylized owl (circa 1906).

Source: Österreichisches Volkshochschularchiv (ÖVA), Fotoarchiv. B-Foto 3/251/–500/323.

The founding appeal states that in the interaction of "all corporations", a "concentration of the various educational efforts" should be created in one place. In continuation of the popular university lectures established at the University of Vienna according to the English model, and following the model of the French *Université Populaire*, more in-depth study of the subjects offered was to take place through "guided reading and special courses" and "personal contact between learners and teachers" (ibid., 1–2). To assure such contact, reading, library, and lecture rooms were to be united at this "concentration point", with rooms "in which musical productions or physical demonstrations take place or travelling exhibitions of works of art were to be organized [*sic*]". Other charitable institutions of various kinds were asked to join, until the *Volksheim*, the popular university (*Volkshochschule*), would be transformed into a real "people's palace" (ibid.).

The new association, which sought to enshrine the name *Volkshoch-schule* in the statutes as a model of a popular university, had to acquire a less suspicious-sounding association name after an official objection from the Lower Austrian governor's office. The conceptual connection between *Volk* ("People") and *Hochschule* (the generic term for higher education institutions) seems to have been taken as an affront by the police authority responsible for admitting or disallowing associations (Stifter 2002, 103). The widely used term *Volksheim* was therefore chosen as an alternative, since it also took into account the social dimension intended from the outset. But as Ludo Hartmann noted in the report on the first general assembly of the young association: "'Volksheim' is written and 'Volks-Universität' is spoken" (Das Volksheim 1901, 1). The aim was to create a place of learning "without social partitions . . . in which members of the most diverse social strata could interact without bias" (Rothe 1906, 1). In addition to the curriculum on offer in the narrower sense, this program was also a matter of achieving the intended "intellectual and social rapprochement between the various classes of the population" (Aufruf zur Gründung 1901, 1), in order to correct "the prejudices formed by political life" by means of contacts among different social circles, as the secretary Josef Luitpold Stern wrote in 1910 (Rothe 1906, 5–6).

Located on the outskirts of the city, in the workers' district of Ottakring, this first urban people's university under the direction of its founding chairman, the mineralogist Friedrich Becke (Tertsch 1956; Filla 1993), already had a lending library at its initial location, in a basement at the heavily frequented Urban-Loritzplatz 1; its 4,000 books had been donated by the economist Otto Wittelshöfer (Ein Abend im Volksheim 1907, 7). In addition to lecture halls, there was a reading room with 35 artistic and scientific journals; an experimental physics laboratory, which, as one proudly noted, even the "first university of the Reich, the University of Vienna, has to do without" (*Österreichische Volkszeitung* 1902); and there were also a chemical as well as a photographic laboratory. The whitewashed lecture hall was decorated with "Secessionist ornaments in green and dark yellow" and was open to all, to the exclusion of all political aspirations, as chairman Friedrich Becke emphasized in his opening speech: "Whoever wants the truth, come to us, be he a radical or a conservative" (Die Eröffnung des Volksheims 1901, 5). There was also room for self-organization. A kitchenette set up by members of the association made it possible for participants "to eat the food they had brought with them in the kitchen, which had been adapted as a social room, with a warm, alcohol-free drink" (1. Jahresbericht des Vereines 'Volksheim' 1902, 93). In this way, the *Volksheim* was to become a social center without, of course, "having to forgo its scientific character because of this", as the first annual report states. Although the literary scholar Emil Reich, after a study trip to England, Scotland, and France for scientific purposes in June and July 1901, modestly reported "what a long way there is still to go" (Central-Verband 1901, 6) in Austria, the *Börsenblatt für den deutschen Buchhandel* already stated in October 1902 that Vienna now holds "the first rank in the field of popular education" (Schiller 1902).

Due to the high attendance at the daily evening courses, around 600 each in the first year, additional rooms were rented in rapid succession. The enormous influx of some highly renowned lecturers—the list reads today like the Who's Who of the scientific, intellectual, and artistic elite of that time—could not be met by this ad-hoc spatial expansion. Therefore, in 1903, the honorary board of the association, headed by the chairman Friedrich Becke, already planned to erect its own building with adequately equipped rooms for the *Volksuniversität*. In his function as deputy chairman, Ludo Moritz Hartmann succeeded within a short period of time in attracting major industrialists, bankers, and members of the high nobility to the project. For example, Rudolf Auspitz, Margrave Pallavicini, and Philipp von Schoeller donated money to purchase the building site at the corner of Kofler Park and Neumayergasse in the Vienna district of Ottakring. In addition to Karl Wittgenstein and Baron Albert Rothschild, Hartmann succeeded in persuading other members of the Viennese bourgeoisie and Viennese merchants to pay annual building fund contributions to cover the mortgage loan. Friedrich Becke, on the occasion of the laying of the foundation stone for the new building in December 1904, said that the companies involved, such as the Prague Iron Industry Society, Alpine Montan, Wienerberger Ziegelfabrik, or Perlmooser Zementfabrik, supported the completion of the building by providing financial and material support (Die Grundsteinlegung 1904, 8).

After the plans had been drawn up by building inspector Franz von Neumann under the direction of architect Ludwig Faigl, the foundation stone was laid in December 1904. As secretary of the association, Emil Reich read the document, which was solemnly attached to the foundation stone of the house. The text, which probably originates from the worker poet Josef Luitpold Stern, reads in part as follows:

"On 24 February 1901, workers, citizens, and university teachers founded the association 'Volksheim' as a place of higher scientific education and rich artistic enjoyment for the broad strata of working people. . . . As far as the Volksbildungsverein and popular university courses were concerned, numerous courses and laboratories were held here. In strict impartiality, far removed from the disputes of the day, the Volksheim imparts to every striving the assured knowledge of science and the rich, full creations of art. The building that we . . . have erected in Vienna's workers' quarter . . . is intended to serve these purposes. In this 'Volksheim' building, as a home of the freely accessible people's university, may the rays of light illuminate the heads and warm the hearts; may it mean a deeper and more elevated life for all those who strive upwards, and may it serve the best hopes of the future" (ibid., 9).

In his capacity as chairman of the Vienna Popular Education Association, Friedrich Jodl concluded his congratulatory speech with the wish "that the spirit of unity between science and the people, which forms the unchanging basis for the future, should prevail in the new house" (ibid.).

As the *Neues Wiener Tagblatt* underlined, the speeches by university professors at the opening made it clear that "modern science does not wear wigs, that it does not isolate itself from the people in cool, unworldly nobility" (*Neues Wiener Tagblatt* 1904, 1). And the *Wiener Allgemeine Zeitung* reported that the meeting, despite the presence of individual "radical politicians", primarily had a "scholarly character"; "excellent teachers of science and world famous researchers" had united to form this work, which "in the midst of the hideous political noise, the raging insults, the domination of slogans, and the total fissure and dissension of the fatherland, rises like a quiet island" (Das Volksheim 1904). The *Neues Wiener Tagblatt* also emphasized that the longing for knowledge was by no means a "transcendental feeling", but was based on "practical, even real knowledge" and on the realization that "the knowing person sees sharper, hears better, works more expediently and, above all, lives healthier than the ignorant person" (*Neues Wiener Tagblatt* 1904, 1).

The building, the design and "solidity of execution" of which (Arlt 1905) attracted a great deal of public attention, was opened just under a year later, on November 5, 1905. Among the numerous keynote speakers were the Dean of the Philosophical Faculty of the University of Vienna, the President of the Lower Austrian Chamber of Commerce and Industry, and professors from the Universities of Innsbruck and Prague. Also present were the President of the Academy of Sciences, Eduard Suess, representatives of the Ministry of Finance, the Ministry of Commerce, and the University of Natural Resources and Applied Life Sciences.

The learning locations, furnished with well-equipped laboratories, spacious lecture theatres, scientific cabinets, bright reading rooms, and extensive specialist libraries, as well as the possibility of observing the heavens, were geared to the ideal of an academic forum, free of domination and ideology, whose primary goal was peaceful communication and the individual and collective further development of all involved. The building itself, a three-storey corner house with a large window front, built on the foundations of a former sand pit, contained a 500-person amphitheatre-like hall with a gallery in the central interior, the furnishings of which had been donated by Baron Albert Rothschild. Equipped with a mobile experiment desk including a gas extraction hose, gas lighting, and water taps, as well as connections for high-voltage current, chemical and physical experiments could be carried out in front of large audiences; in addition, the hall provided a venue for concerts and theatre performances.

The second-largest lecture hall had 200 seats, followed by a lecture hall for chemistry and physics, a natural history lecture hall, and a number of smaller lecture halls. In addition, there was a chemical laboratory with 24 workstations, also frequented by university students, and two adjoining rooms for quantitative and synthetic-preparative work, as well as a fume hood with a shower for extinguishing clothing that had caught fire; a physical laboratory with a collection of apparatus and DC three-phase current converters, and also a natural history cabinet with preparation room and zoological, botanical, and mineralogical collections (Figure 10.2).

There was also an experimental-psychological cabinet, an aquarium and terrarium room, an art-historical cabinet, a large drawing room, several studios for photographic work and three darkrooms, a room for experiments with explosive materials, and an acid depot chamber. For gymnastics, there was a gymnasium with cloakroom and shower room in the basement of the building (Faigl 1906, 237), and the park in front of the building had "a botanical garden and playgrounds for all kinds of sports" (Das Volksheim 1901, 1).

Figure 10.2 Natural Science Section (*Fachgruppe*) at microscopy exercises in the Natural History Cabinet of the 'People's University' Ottakring (circa 1910).

Source: Österreichisches Volkshochschularchiv (ÖVA), Fotoarchiv. B-Foto 3/251–500/330.

The well-designed building was illuminated by incandescent gas light, and the large halls by electric light. All rooms were connected to each other and to the central office, where the "state telephone" (*Staatstelephon*) was located (Arlt 1905). On the mezzanine floor of the brightly decorated building, there was a reading room for 140 people and the adjacent public library with 20,000 volumes, as well as the in-house popular science library with 6,000 volumes, made up of donations, whose reading room was connected to the book depot on the mezzanine floor by a book lift. On the third floor, there were also special libraries for so-called *Fachgruppen*—seminar-like working groups of laypersons and experts that had existed for years and had their own statutes as independent sections of the Volksheim (Filla 2001).

In the vestibule, there was a marble plaque which read: "Built in 1905 by the harmonious cooperation of men and women from the working, civic, and university circles" (Bau des Volksheims 1905, 970). The ground floor of the building also housed a canteen run by the "Association of Women Teetotallers"—the only non-alcoholic restaurant in the surrounding area—in order to protect listeners from the "pernicious atmosphere of brandy" (Das Volksheim n.d.), as was already the case at the previous location.

This building, architecturally designed in "noble simplicity" (Bau des Volksheims 1905, 970), gained its particular attractiveness through a sense of community that is hardly comprehensible today, the great continuity of the *Volksheim* lecturers, the rejection of any methodological constraint, the voluntary nature of participation, the atmosphere of collective self-organization and enthusiasm for learning, the coexistence of the most diverse disciplines, and the enormous range of educational offerings. The curriculum ranged from basic subjects and advanced lectures in almost all academic disciplines to artistic events, special courses in individual sciences, and excursions—for example, to art galleries or to the *Wiener Werkstätten*. The participants' own activities were supported by holding evening discussions. On average, 40 to 45 percent of the participants were women. Special attention was paid to banishing the often "obstructive shyness" about asking a question by tactfully answering "every question, even those that are not correctly asked or reveal larger gaps" (Rothe 1906, 4).

The heterotopic dimension inherent in this social interaction had already been expressed years before by Emil Reich in the context of the predecessor institution, the popular university lectures. As he wrote in 1897: "And already this is in such an excited, agitated time an estimable success that members of all classes of society and age groups, men and women without distinction, gather as equal persons in the same room in order to listen together to a quiet, objective, and scientific explanation of things that only concern knowledge of the truth. . . . Their teachers are men of talent and diligence, but men whose simple jackets reveal that their material situation is not so ideal . . . should this not in our materialistic, money-hungry age exert an indirect impact, which may be all the more lasting precisely because it is unintended?" (Reich 1897, 26).

Final Remark

Vienna's adult education centers eventually developed into an academy of popular science, a 'shadow university' on a small scale—albeit without any possibility of obtaining degrees—especially at the time of their heyday in the First Republic. Municipal subsidies and donations from the trade unions and chambers of labor made it possible both to further differentiate the range of subjects on offer and to decentralize the program to the entire city by establishing branch offices. In addition, and in contrast to increasingly anti-Semitic racist activities at the University of Vienna (Ash 2015, 75ff.; Taschwer 2015, Chapter 3), the academic teaching staff increased massively due to the increasing proportion of lecturers with Jewish backgrounds, which exceeded 50 percent by the end of the 1920s (Stifter 2018, 496). In 'Red Vienna', the adult education centers objectively acted as "bastions" of the Enlightenment and as "intellectual armouries" against increasingly conservative Catholic and reactionary cultural and identity politics (Stifter 2019; Schwarz et al. 2019).

And yet, already on the occasion of the tenth anniversary of the founding of the Volkshochschule Ottakring in February 1911 and one year after completion of the building described previously, Ludo Hartmann, the spiritus rector of this avant-garde urban educational institution, thought that the educational vision of the pioneer generation had been achieved: "what appeared to be a utopia at the constituent assembly on 24 February 1901, the construction of a fully equipped adult education centre in Vienna, is today doubly truth" (Hartmann 1911, 2).

In view of the learning and teaching structures outlined here, the social space created at the *Volksheim*, and the novel relationships among the actors involved, the seemingly utopian concept of a scientifically oriented education *for all* as well as the specific *exclusivity* of those places of education, which also manifested itself in a separate "supervisory service" of members who ensured that "no unauthorised person penetrates" (Siemering 1911, 33), the following can still be said: The places of Viennese popular education described in this chapter can already at the time of the monarchy be regarded as educational counter-places, as "locally realised utopias" in the sense of Foucault's heterotopology.

Bibliography

Archival Sources

Österreichisches Volkshochschularchiv (ÖVA), Bestand "Volksheim Ottakring."
Österreichisches Volkshochschularchiv (ÖVA), Bestand "Volksheim Ottakring," Zeitungsausschnitte-Sammlung, 1901–1912.

216   *Christian H. Stifter*

Primary and Secondary Literature

1. Jahresbericht des Vereines 'Volksheim' in Wien. *Zentralblatt für Volksbildung-swesen* 2 (1902), April, 6–7, 81–94.

Altenhuber, H., and Pfniß, A. (eds.) (1965). *Bildung—Freiheit—Fortschritt. Gedanken österreichischer Volksbildner.* Vienna: Verband Österreichischer Volkshochschulen.

Arlt, F. V. (1905). Das neue Volksheim in Wien. *Grazer Tagespost,* Novrefember 5 1905. Quoted from: ÖVA, Bestand "Volksheim Ottakring," Zeitungsauss-chnitte-Sammlung, 1901–1912.

Ash, M. G. (2015). Die Universität Wien in den politischen Umbrüchen des 19. und 20. Jahrhunderts. In: Ash, M. G., and Ehmer, J (eds.), *Universität—Politik—Gesellschaft. 650 Jahre Universität Wien—Aufbruch ins neue Jahrhundert. Bd. 2.* Göttingen: V & R Unipress, 29–172.

Ash, M. G., and Stifter, Ch. H. (eds.) (2002). *Wissenschaft, Politik und Öffentlichkeit. Von der Wiener Moderne bis zur Gegenwart.* Vienna: WUV Universitätsverlag.

Aufruf zur Gründung eines Volksheims (Volkshochschule) (1901), 3. ÖVA, Bestand "Volksheim Ottakring," Zeitungsausschnitte-Sammlung, 1901–1912.

Bau des Volksheims (1905). *Der Bautechniker. Zentralorgan für das österreichische Bauwesen* 25:45, November 10, 970.

Bauer, O. (1906). Die Wiener Arbeiterschule. 1905/1906. *Neue Zeit* 24 (1906) 2, 462. Quoted from Siemering, H. (1911). *Arbeiterbildungswesen in Wien und Berlin. Eine kritische Untersuchung.* Karlsruhe: Braunsche Hofdruckerei und Verlag, 7.

Central-Verband der deutsch-österreichischen Volksbildungs-Vereine. IV. ordentlicher Delegiertentag, abgehalten am November 2, 1901. ÖVA, Bestand "Volksheim Ottakring."

Das Volksheim. Bericht über die constituirende Generalversammlung vom 24. Februar 1901. *Das Wissen für alle. Volksthümliche Vorträge und Populär Wissenschaftliche Rundschau* 10, March 3, 1.

Das Volksheim (1904). *Wiener allgemeine Zeitung,* 25, December 20. Quoted from ÖVA, Bestand "Volksheim Ottakring," Zeitungsausschnitte-Sammlung, 1901–1912.

Das Volksheim (n.d.). Quoted from ÖVA, Bestand "Volksheim Ottakring," Zeitungsausschnitte-Sammlung, 1901–1912.

Defert, D. (2017). Raum zum Hören. In: Foucault, M. (ed.), *Die Heterotopien. Der utopische Körper. Zwei Radiovorträge.* Frankfurt am Main: Suhrkamp, 69–92.

Der Wiener Frauenclub. Eine gesellschaftliche Action. *Neues Wiener Tagblatt* 34 (1900) 106, April 19, 4.

Die Eröffnung des Volksheims. *Arbeiter-Zeitung* 13 (1901) 113, April 26, 5.

Die Grundsteinlegung zum Neuen 'Volksheim'. *Neue Freie Presse*: 14484 (1904) December 19, 8.

Ehalt, H. Ch. (1978). Das Wiener Schulwesen in der liberalen Ära. In: Czeike, F. (Ed.), *Wien in der liberalen Ära. Forschungen und Beiträge zur Wiener Stadtgeschichte. Bd. 1. Sonderreihe der Wiener Geschichtsblätter.* Vienna: Deuticke, 120–147.

Ein Abend im Volksheim. *Arbeiter-Zeitung* 19 (1907) 314, November 15, 7.

Engelbrecht, H. (1984). *Geschichte des österreichischen Bildungswesens. Erziehung und Unterricht auf dem Boden Österreichs. Bd. 3. Von der frühen Aufklärung bis zum Vormärz*, Österreichischer Bundesverlag.

Faigl, L. (1906). Das Deutsche 'Volksheim' in Wien, XVI. Bezirk. *Der Bautechniker. Zentralorgan für das österreichische Bauwesen* 26 (1906) 12, März 23, 237–240.

Feichtinger, J. (2010). *Wissenschaft als reflexives Projekt. Von Bolzano über Freud zu Kelsen. Österreichische Wissenschaftsgeschichte 1848–1938*. Bielefeld: Transcript.

Filla, F. (2001). *Wissenschaft für alle—ein Widerspruch? Bevölkerungsnaher Wissenstransfer in der Wiener Moderne. Ein historisches Volkshochschulmodell* (Schriftenreihe des Verbandes Österreichischer Volkshochschulen, Bd. 11). Innsbruck, Vienna, Munich: Studien-Verlag.

Filla, W. (1993). Weltbekannter Mineraloge und Volksbildner. Ein Kurzportrait Friedrich Beckes (1855–1931). *Spurensuche. Zeitschrift für Geschichte der Erwachsenenbildung und Wissenschaftspopularisierung* 4:1, 17–23.

Filla, W. (1994). Arbeiter als Teilnehmer in den Wiener Volkshochschulen der zwanziger Jahre. In: Stifter, Ch. (Ed.), *Arbeiterbildung und Volkshochschule. Von den Anfängen bis in die Zwischenkriegszeit. Schweiz—Deutschland—Österreich. Tagungsbericht*. Vienna: Verein zur Geschichte der Volkshochschulen, 1–8.

Filla, W. (2001). Miss A. S. Levetus—Kunsthistorikerin und Volksbildnerin. Porträt einer grenzüberschreitenden Pionierin. *Spurensuche. Zeitschrift für Geschichte der Erwachsenenbildung und Wissenschaftspopularisierung* 12:1–4, 24–39.

Filla, W., Judy. M., and Knittler-Lux, U. (eds.) (1992). *Aufklärer und Organisator. Der Wissenschaftler, Volksbildner und Politiker Ludo Moritz Hartmann* (Schriftenreihe des Verbandes Wiener Volksbildung 17). Vienna: Picus Verlag.

Foucault, M. (2017). *Die Heterotopien. Der utopische Körper. Zwei Radiovorträge*. Frankfurt am Main: Suhrkamp.

Frank, H. (1970). Friedrich Jodl (1849–1914). Seine Lehre und seine Rolle in der bürgerlichen Reformbewegung Österreichs und Deutschlands. Phil. Diss., Universität Freiburg i. Br.

Ganglbauer, St. (1999). 'Neutrale' Volksbildung und die 'wertungsfreie Wissenschaft'. Die 'Sehnsucht nach Schicksal und Tiefe' und der Richtungsstreit in der deutschsprachigen Volksbildungsbewegung der 20er-Jahre. *Spurensuche. Zeitschrift für Geschichte der Erwachsenenbildung und Wissenschaftspopularisierung* 10:1–4, 60–84.

Hartmann, L. M. (1901). Volksbildung und Ethik. Offener Brief an Herrn Dr. F. W. Förster. 1901. *Die Zeit* 350 (1901) Juni 15, 165. Quoted from Filla, W. (1992). Ludo Moritz Hartmann. Wissenschaftler in der Volksbildung. In: Filla, W., Judy. M., and Knittler-Lux, U. (eds.), *Aufklärer und Organisator. Der Wissenschaftler, Volksbildner und Politiker Ludo Moritz Hartmann* (Schriftenreihe des Verbandes Wiener Volksbildung 17). Vienna: Picus Verlag, 84.

Hartmann, L. M. (1910). Das Volkshochschulwesen. Seine Praxis und Entwicklung nach Erfahrungen im Wiener Volksbildungswesen. In: Altenhuber, H., and Pfniß, A. (eds.), *Bildung -Freiheit—Fortschritt. Fortschritt. Gedanken österreichischer Volksbildner*. Vienna: Verband Österreichischer Volkshochschulen 1965, 115–130.

218   *Christian H. Stifter*

Hartmann, L. M. (1911). Das Wiener Volksheim, in: *Arbeiter-Zeitung*, 23: 57, February 26, 2.

Hartmann, L. M. (1920). Demokratie und Volksbildung. In: Altenhuber, H., and Pfniß, A. (Eds.), *Bildung—Freiheit—Fortschritt. Gedanken österreichischer Volksbildner*. Vienna: Verband Österreichischer Volkshochschulen, 131–134.

Hartmann, L. M. (1921). Ein Kulturjubiläum. Zum zwanzigsten Geburtstag des Ottakringer Volksheims. *Arbeiter-Zeitung* 23:111, April 24, 2.

Jodl, F. (1911). Eröffnung des Volksbildungshauses Margareten. *Neue Freie Presse*, Nr. 16675, January 23, 4.

Jodl, F. (1912). Geleitwort. *25 Jahre Volksbildung. Chronik des Wiener Volksbildungsvereines von 1887 bis 1912*. Vienna: Verlag des Wiener Volksbildungsvereines, 1.

Kapner, G. (1961). *Die Erwachsenenbildung um die Jahrhundertwende, dargestellt am Beispiel Wiens*. Vienna: Notring.

Kniefacz, K., Nemeth, E., Posch, H., and Stadler, F. (eds.) (2015). *Universität—Forschung—Lehre. Themen und Perspektiven im langen 20. Jahrhundert. 650 Jahre Universität Wien—Aufbruch ins neue Jahrhundert, Bd. 1*. Göttingen: V & R Unipress.

Large, J. (2003). The Studio and the Workshops: Amelia Levetus and the British Influence on the Applied Arts in Vienna. Thesis, Central European University Budapest.

Lefebvre, H. (1991). *The Production of Space*. Oxford: Blackwell.

Levetus, A. S. (1905). *Imperial Vienna: An Account of Its History, Traditions and Arts*. London and New York: John Lane.

Liebknecht, W. (2012). *Kleine Politische Schriften*. Berlin: Tredition.

Mach, E. (1896). *Populärwissenschaftliche Vorlesungen*. Leipzig: Verlag Johann Ambrosius Barth.

Neues Wiener Tagblatt. Demokratisches Organ 38 (1904) December 20, 1.

Österreichische Volkszeitung (1902) May 7. Quoted from ÖVA, Bestand "Volksheim Ottakring," Zeitungsausschnitte-Sammlung, 1901–1912.

Rathkolb, O. (2013). Gewalt und Antisemitismus an der Universität Wien und die Badeni-Krise 1897. Davor und danach. In: idem. (ed.), *Der lange Schatten des Antisemitismus. Kritische Auseinandersetzungen mit der Geschichte der Universität Wien im 19. und 20. Jahrhundert*. Göttingen: V & R unipress, 69–92.

Reich, E. (1897). Volkstümliche Universitätsbewegung. In: Schweizerischen Gesellschaft für ethische Kultur (Ed.), *Ethisch-socialwissenschaftliche Vortragskurse, veranstaltet von den ethischen Gesellschaften in Deutschland, Oesterreich und der Schweiz* (Züricher Reden, Bd. 5). Bern: Steiger & Cie.

Reichspost (1900). Liberale Volksverbildung. *Reichspost. Unabhängiges Tagblatt für das christliche Volk Oesterreich-Ungarns* 7:50, March 3, 54.

Rothe, K. C. (1906). Das Volksheim in Wien. Bericht über die Tätigkeit des Vereins 'Volksheim' und Schilderung des am 5. November 1905 eröffneten eigenen Hauses. *Concordia. Zeitschrift der Centralstelle für Arbeiter-Wohlfahrtseinrichtungen* 7 (1906). ÖVA, Bestand "Volksheim Ottakring," Zeitungsausschnitte-Sammlung, 1901–1912.

Rumpler, H., and Seeger, M. (2010). *Die Gesellschaft der Habsburgermonarchie im Kartenbild. Verwaltungs-, Sozial- und Infrastrukturen. Nach dem Zensus von 1910* (Die Habsburgermonarchie 1848–1918, Vol. 9/2). Vienna: Verlag der Österreichischen Akademie der Wissenschaften.

Schiller, F. (1902). Wiener Brief. *Sonderabdruck aus dem Börsenblatt für den Deutschen Buchhandel*, 254 (1902) November 1. Quoted from ÖVA, Bestand "Volksheim Ottakring," Zeitungsausschnitte-Sammlung, 1901–1912.

Schwarz, W. M., Spitaler, G., and Wikidal, E. (eds.) (2019). *Das Rote Wien 1919–1934. Ideen—Debatten—Praxis* (Katalog zur 426. Sonderausstellung des Wien Museums). Basel: Birkhäuser.

Seebacher, F. (2010). *Das Fremde im 'deutschen' Tempel der Wissenschaften. Brüche in der Wissenschaftskultur der Medizinischen Fakultät der Universität Wien.* Vienna: Verlag der Österreichischen Akademie der Wissenschaften.

Siemering, H. (1911). *Arbeiterbildungswesen in Wien und Berlin. Eine kritische Untersuchung.* Karlsruhe: Braunsche Hofdruckerei und Verlag.

Sport & Salon. Illustrierte Zeitschrift für die vornehme Welt 8 (1905) 16, April 22, 9.

Stadler, F. (1997). *Studien zum Wiener Kreis. Ursprung, Entwicklung und Wirkung des Logischen Empirismus im Kontext.* Frankfurt am Main: Suhrkamp.

Stifter, Ch. (ed.) (1994). *Arbeiterbildung und Volkshochschule. Von den Anfängen bis in die Zwischenkriegszeit. Schweiz—Deutschland—Österreich. Tagungsbericht.* Vienna: Verein zur Geschichte der Volkshochschulen.

Stifter, Ch. H. (2002). Die Wiener Volkshochschulbewegung in den Jahren 1887–1938: Anspruch und Wirklichkeit. In: Ash, M. G., and Stifter, Ch. H. (eds.), *Wissenschaft, Politik und Öffentlichkeit. Von der Wiener Moderne bis zur Gegenwart.* Vienna: WUV Universitätsverlag, 95–116.

Stifter, Ch. H. (2005). *Geistige Stadterweiterung. Eine kurze Geschichte der Wiener Volkshochschulen, 1887–2005.* Weitra: Bibliothek der Provinz.

Stifter, Ch. H. (2015a). Ludo Moritz Hartmann (1865–1924). Wissenschaftlicher Volksbildner—sozialdeterministischer Historiker—realitätsferner Politiker. In: Ash, M. G., and Ehmer, J. (eds.), *Universität—Politik—Gesellschaft. 650 Jahre Universität Wien—Aufbruch ins neue Jahrhundert. Bd. 2.* Göttingen: VR Unipress, 247–263.

Stifter, Ch. H. (2015b). Universität, Volksbildung und Moderne—die 'Wiener Richtung' wissenschaftsorientierter Bildungsarbeit. In: Kniefacz, K., Nemeth, E., Posch, H., and Stadler, F. (eds.), *Universität—Forschung—Lehre. Themen und Perspektiven im langen 20. Jahrhundert. 650 Jahre Universität Wien -Aufbruch ins neue Jahrhundert* (Vol. 1). Göttingen: V & R-Unipress, 293–316.

Stifter, Ch. H. (2018). Antisemitismus und Volksbildung vor 1938—ein Ausschlussverhältnis? In: Enderle-Burcel, G., and Reiter-Zatloukal, I. (eds.), *Antisemitismus in Österreich 1933–1938.* Vienna: Böhlau Verlag, 487–508.

Stifter, Ch. H. (2019). Volkshochschulen im Roten Wien. In: Schwarz, W. M., Spitaler, G., and Wikidal, E. (eds.), *Das Rote Wien 1919–1934. Ideen—Debatten—Praxis* (Katalog zur 426. Sonderausstellung des Wien Museums). Basel: Birkhäuser, 114–119.

Taschwer, K. (2002). Wissenschaft für viele. Zur Wissenschaftsvermittlung im Rahmen der Wiener Volksbildung um 1900. PhD dissertation, University of Vienna.

Taschwer, K. (2015). *Hochburg des Antisemitismus. Der Niedergang der Universität Wien im 20. Jahrhundert.* Vienna: Czernin.

Tertsch, H. (1956). Erinnerungen an Friedrich Becke. In: *Mitteilungen der Österreichischen Mineralogischen Gesellschaft* (Vol. 4). Vienna: Sonderdruck.

Verein Ernst Mach (ed.) (1929). *Wissenschaftliche Weltauffassung. Der Wiener Kreis. Veröffentlichungen des Vereines Ernst Mach*. Vienna: Artur Wolf Verlag. Quoted from Stöltzner, M., and Uebel, Th. (eds.) (2006). *Wiener Kreis. Texte zur wissenschaftlichen Weltauffassung von Rudolf Carnap, Otto Neurath, Moritz Schlick, Philipp Frank, Hans Hahn, Karl Menger, Edgar Zilsel und Gustav Bergmann*. Hamburg: Felix Meiner Verlag, 3–29.

Verein zur Abhaltung akademischer Vorträge für Damen. *Das Vaterland*, 37:169 (1896), June 20, 3.

Vortrag einer jungen Dame an der Universität Wien. *Neue Freie Presse*, No. 11655 (1897), February 2, 6.

Whitehead, A. N. (1988). *Wissenschaft und moderne Welt*. Frankfurt am Main: Suhrkamp.

Contributors

Mitchell G. Ash is Professor Emeritus of Modern History at the University of Vienna, Austria, and a member of the Berlin-Brandenburg Academy of Sciences and Humanities as well as the European Academy of Sciences and Arts. He has published widely on the political, social, and cultural relations of the sciences in the nineteenth and twentieth centuries. Relevant publications include: *The Nationalization of Scientific Knowledge in the Habsburg Empire (1848–1918)* (ed. with Jan Surman). Basingstoke: Palgrave Macmillan 2012; Multiple Modernisms in Concert: The sciences, technology and culture in Vienna around 1900. In: Bud, R. et al. (eds.), *Being Modern: The Cultural Impact of Science in the Early Twentieth Century*. London: University College London Press 2018, 23–39.

Sándor Békési is a curator at the Vienna Museum in the Department of City Development and Topography. He is author of various publications on urban history, environmental and transport history, most recently Sándor Békési and Elke Doppler (eds.), *Wien von oben. Die Stadt auf einen Blick*. Vienna: Metroverlag 2017, and Wem gehört(e) der Straßenraum? Zur Sozialgeschichte urbaner Mobilität. In: Assmann, A., Assmann, J., and Rathkolb, O. (eds.), *Geschichte und Gerechtigkeit*. Munich: LIT 2019, 34–40.

Dorothee Brantz is Professor of Urban Environmental History and Director of the Center for Metropolitan Studies at the Technical University of Berlin. Her research centers on the relations between human, animals, and urban environments in metropolitan settings from the nineteenth century to the present. Her most recent publications include: *Greening the City: Urban Landscapes in the Twentieth Century* (co-ed. with Sonja Dümpelmann). Charlottesville: University of Virginia Press, PB 2019, and *100 Jahre Groß-Berlin: Grünfrage und Stadtentwicklung* (co-ed. with Harald Bodenschatz). Berlin: Lukas 2019.

Oliver Hochadel is Tenured Scientist at the Institutció Millà i Fontnanals—CSIC (Spanish National Research Council) in Barcelona. He has published widely on the history of science and its publics from the eighteenth to the twenty-first centuries. Recent book publications include: *Barcelona: An Urban History of Science and Modernity, 1888–1929* (ed. with Augustí Nieto-Galán). London and New York: Routledge 2016; *Urban Histories of Science: Making Knowledge in the City 1820–1940* (ed. with Augustí Nieto-Galán). London: Routledge 2018; and Interurban Knowledge Exchange in Southern and Eastern Europe, 1870–1950 (ed. with Eszter Gantner and Heidi Hein-Kircher). London: Routledge, 2020.

Marianne Klemun is Professor at the Department of Modern History at the University of Vienna since 2002. Her research fields include cultures, practices, and political contexts of the history of natural history (geology, botany, and gardens). Recently published books are: *Nicolaus Jacquin: Ein Naturforscher (er)findet sich* (with Helga Hühnel). Göttingen: VR unipress 2017; *Expeditions as Experiments: Practicing Observation and Documentation* (ed. with Ulrike Spring). Basingstoke: Palgrave Macmillan 2016; Skulls and Blossoms: Natural History Collections and Their Meanings. Special issue of *Centaurus. An International Journal of the History of Science and Its Cultural Aspects*, 60:4 (2018), printed in 2019, ed. with Anastasia Fedotova and Marina Loskutova; *Wissenschaft als Kommunikation in der Metropole Wien. Die Tagebücher Franz von Hauers der Jahre 1860–1868*. Sole editor. Vienna, Cologne, Weimar: Böhlau to appear 2020.

Johannes Mattes is a postdoctoral scholar at the Austrian Academy of Sciences and lecturer at the Department of History at the University of Vienna. His current research examines the history of natural sciences (especially speleology, earth sciences, geography, and cartography), scholarly societies, popular science, and the relationship between politics and research. His most recent publications are: *Reisen ins Unterirdische. Eine Kulturgeschichte der Höhlenforschung*. Vienna, Cologne and Weimar: Böhlau 2015; and *Wissenskulturen des Subterranen. Vermittler im Spannungsfeld zwischen Wissenschaft und Öffentlichkeit. Ein biografisches Lexikon*. Vienna: Böhlau 2019.

Brooke Penaloza-Patzak is Erwin Schrödinger Postdoctoral Fellow at the Department for the History and Sociology of Science, University of Pennsylvania, and the Department of Economic and Social History, University of Vienna. Her current research areas are the transnational movements of researchers and research objects in cultural anthropology and the history of science and scholarship on the Bering Straight during the nineteenth and twentieth centuries. Recent publications include: An Emissary From Berlin: Franz Boas and the Smithsonian Institution, 1887. *Museum Anthropology* 41:1 (2018), 30–45; and Das Emergency Society for German and Austrian Science and Art, 1920–1927. Eine

Einführung in eine beinahe unbekannte Hilfsorganisation und der Mehrwert ihrer Erforschung. In: J. Feichtinger, M. Klemun, J. Surman, and P. Svatek (eds.), *Wandlungen und Brüche. Wissenschaftsgeschichte als politische Geschichte*. Göttingen: VR unipress 2018, 125–132.

Ulrike Spring is Associate Professor of Modern European History, Department of Archaeology, Conservation and History, University of Oslo. Her research interests include nineteenth-century Arctic history, in particular expeditions and tourism, and public history, with a focus on literary museums. She was leader and co-coordinator of the research project TRAUM—Transforming Author Museum (Research Council of Norway 2016 to 2019). Relevant recent publications include: *Expeditions as Experiments: Practicing Observation and Documentation* (ed. with Marianne Klemun). Basingstoke: Palgrave Macmillan 2016; Cruise Tourists in Spitsbergen Around 1900: Between Observation and Transformation. Nordlit 2020. doi 10.7557/13.5026.

Christian H. Stifter is Director of the Austrian Archives for Adult Education, and editor of the journal *Spurensuche*, which specializes in the history of adult education and popular science. He has been awarded the Innovation Prize of the German Institute for Adult Education (Bonn), the Leibniz Society for the Online Historiography of Adult Education (2008), as well as the Ludo Moritz Hartmann Prize of the Association of Austrian Adult Education Centres (2008). His most recent book is *Zwischen geistiger Erneuerung und Restauration. US-amerikanische Planungen zur Entnazifizierung und demokratischen Neuorientierung österreichischer Wissenschaft 1941–1955*, Vienna: Böhlau Verlag 2014.

Petra Svatek was Research Associate at the Department of History at the University of Vienna from 2006 to 2016. In 2017/2018 she researched the history of geography and cartography in Vienna during the Nazi era with the support of the Future Fund of the Republic of Austria. Since 2018 she is affiliated with the Woldan Collection of the Austrian Academy of Sciences in Vienna. Her main research areas are history of geography, spatial research, and cartography, with emphasis on the connection between science and politics, in Austria and Germany between 1880 and 1950. Recent publications include: *Wissenschftliche Forschung in Österreich 1800–1900* (ed. with Christine Ottner and Gerhard Holzer). Göttingen: VR unipress 2015; and Ethnic Cartography and Politics in Vienna 1918–1945. *British Journal for the History of Science* 51:1 (2018), 91–121.

Index

For Product Safety Concerns and Information please contact our EU
representative GPSR@taylorandfrancis.com
Taylor & Francis Verlag GmbH, Kaufingerstraße 24, 80331 München, Germany